BIOCHEMISTRY RESEARCH TRENDS

ACETIC ACIDS

CHEMICAL PROPERTIES, PRODUCTION AND APPLICATIONS

BIOCHEMISTRY RESEARCH TRENDS

Additional books in this series can be found on Nova's website under the Series tab.

Additional e-books in this series can be found on Nova's website under the e-book tab.

BIOCHEMISTRY RESEARCH TRENDS

ACETIC ACIDS

CHEMICAL PROPERTIES, PRODUCTION AND APPLICATIONS

ANGELO BASILE, Ph.D.
EDITOR

New York

For permission to use material from this book please contact us:
Telephone 631-231-7269; Fax 631-231-8175
Web Site: http://www.novapublishers.com

Library of Congress Cataloging-in-Publication Data

ISBN: 978-1-62948-217-0

Library of Congress Control Number: 2013949352

Published by Nova Science Publishers, Inc. † New York

Contents

Preface

The purpose of this book, containing a small number of chapters but highly selected, is to provide some important aspects of acetic acid especially for what concerns its use and applications. Acetic acid is a renewable source and can be easily obtained by fermentation of biomass. Unlike to the other bio-fuel such as methanol and ethanol, acetic acid is not inflammable, therefore it represents a safe hydrogen carrier. Anyhow, these and many other aspects are deeply described in this book. In the following a short description of each chapter is presented. Chapter 1 (Mapesa and Mahungu) introduces the chemistry as well as the properties of acetic acid, the simplest carboxylic acid after formic acid. The most important aspects of its production are discussed, paying particular attention to: a) oxidation of acetaldehyde to acetic acid; b) methanol carbonylation to acetic acid; c) acetic anhydride; d) vinyl acetate; e) sodium acetate; f) acetate ester. Concerning the commercial vinegars, the chapter also deals with the fermentation process of ethanol by acetic acid bacteria. The various chemical processes involving partial oxidation of ethanol to acetaldehyde, which is eventually converted to acetic acid, are also discussed in detail. Finally, the chapter introduces the following acetic acid applications: acetic acid as a solvent, acetic acid in meat preservation, medicinal effects of acetic acid, health and safety considerations. Chapter 2 (Sarropoulou, Dimassi-Theriou and Therios), after an extensive introduction on various aspects of acetic acid (nomenclature, history, chemical properties and reactions, production and uses), shows in detail the properties of auxins, an important category of acetic acids influencing a number of physiological and developmental processes in plants. The effect of the four auxins (IAA, IBA, NAA and 2,4-D), which are involved in the photosynthetic apparatus affecting leaf chlorophyll content and participating on carbohydrate biosynthesis and metabolism, as well as in proline accumulation in both leaves and roots, are analyzed and discussed. Chapter 3 (Romano, Pergomet, Zinczuk and Brandán) presents a complete characterization of the synthesis, structural and vibrational properties of three quinoline acetic acid derivatives [2-(quinolin-8-yloxy), 2-(5-chloroquinolin-8-yloxy) and 2-(5,7-dichloroquinolin-8-yloxy)] with potentials biological activities. To be also said that the knowledge of the structural and vibrational properties is considered very important for the design and synthesis of new and better quinolinic drugs. In Chapter 4 (Pizzio, Romanelli and Blanco), the Authors review the performance of different recyclable solid acids synthesized in their laboratory as catalysts for the use of acetic acid as acetylating agent. In particular, they find that the alcohols and phenols acetylation with the borated zirconia-based catalysts give very good yields to the desired products with very high selectivity. They have the advantages of giving very good yields and excellent selectivity, at reasonable reaction time, thus leading to low environmental impact, both due to the recyclable solid acids and the clean acetylation agent. Chapter 5 (Iulianelli, Liguori, Dalena and Basile) deals with the aspects of hydrogen production from acetic acid. In their chapter the Authors report that only a few manuscripts deal with conventional reactors (usually operating at T > 500 °C), where the acetic acid steam reforming reaction is carried out using different catalysts producing a hydrogen rich-gas stream containing secondary products such as CO_2, CO, CH_4, etc. With the purpose of producing a CO_x-free hydrogen, to be used for example in PEM fuel cells, this chapter introduces the use of the membrane reactor technology for carrying out the acetic acid steam reforming reaction in only one step-whereas the conventional reactors require further stages for hydrogen purification such as water gas shift reactors (both high and low temperatures), partial oxidation reactors and pressure swing adsorption devices. Various interesting aspects of membrane reactors are discussed in detail. Chapter 6 (Bom, Kochhann, Rodrigues and Fauth) introduces the factors that influence the chemical process of disaggregating carbonate rock. The various disaggregation techniques are revised, considering the immersion of carbonate rocks in acetic acid and ranging kinetic parameters. The usefulness of acetic acid in recovering calcareous and siliceous microfossils from carbonate rocks is presented and discussed. In conclusion, the acetic acid methods are easy to work, also permitting reutilization by distillation of residues for further reactions. Chapter 7 (Ačkar, Šubarić, Jozinović and Babić) deals with the use of acetic acid in food preparation/production. In this chapter the current knowledge on acetic acid and its derivatives application in food industry are summarised paying particular attention also to the nutritional and health aspects. In Chapter 8 (Tanino), the coproduction of vinegar and electricity by fermentation is illustrated in detail. This novel concept has been developed by the application of microbial fuel cell technology to acetic acid fermentation. Acetic acid fermentation and microbial fuel cell technology share the same principle, that is microorganisms are used as the catalyst for the oxidation of an organic compound. The most important aspects of this technology are firstly described. Then, the advantages and the Author's recent study on the coproduction of acetic acid and electricity are shown.

In: Acetic Acids
Editor: Angelo Basile

ISBN: 978-1-62948-217-0

Chapter 1

Chemistry and Industrial Use of Acetic Acid

Job O. Mapesa[1,3,*] ***and Symon M. Mahungu***[2,†]
[1]Kenya Methodist University, Department of Human Nutrition and Dietetics, Nairobi, Kenya
[2]Egerton University, Department of Dairy and Food Science and Technology, Egerton, Kenya
[3]Agri-Food and Nutrition Innovations, Nairobi, Kenya

Abstract

Glacial (concentrated) acetic acid also known as ethanoic acid is a carboxylic compound responsible for the characteristic odor and sour taste of vinegar, which means "sour wine" in French. Formation of vinegar is through the oxidative action of aerobic bacteria on dilute solutions of alcohols including wines, malt or cider. This chemical change is the reaction of ethanol with oxygen to form acetic acid and water. Vinegar is about 4-18 % acetic acid. The chemistry and behavior of acetic acid is well defined. Pure acetic acid is classified as a weak acid, clear, colorless liquid with a sharp, irritating odor, with a freezing point below 16 °C and a boiling point of 118 °C.

It is also miscible with water, ethyl alcohol and diethyl ether due to hydrogen bonding properties. Due to its acidic nature, vinegar has acquired a variety of uses industrially and around the household. It is used as a solvent for mineral deposits and organic compounds, flavoring agent in cooking and salad dressings, food preservative, and disinfectant in cleaning agents.

It is also used in industrial manufacture of acetate esters including cellulose acetate in textile industry and vinyl acetate. Use of acetic acid is gaining popularity in the agricultural sector as a fungicide and herbicide and in the production of pharmaceuticals

* Job O. Mapesa: Kenya Methodist University, Department of Human Nutrition and Dietetics, P.O Box 45240-00100, Nairobi, Kenya. Agri-Food and Nutrition Innovations, P.O Box 36374, 00200 Nairobi, Kenya. E-mail: jmapesa@yahoo.com.
† Symon M. Mahungu: Egerton University, Department of Dairy and Food Science and Technology, P.O Box 536, Egerton, Kenya. E-mail: smahungu@yahoo.com.

such as aspirin using acetic anhydride. The aim of this chapter is to give insights into the chemical properties, production and applications of acetic acid.

We will focus on the developments in the chemistry and industrial use of acetic acid or vinegar (acetic acid and vinegar will be used interchangeably) in the last two decades, as well as draw attention to new emerging areas in acetic acid research.

Introduction

Acetic or ethanoic acid (molecular formula, CH_3COOH) is a clear, colorless liquid with a pungent odor found in vinegar at a concentration of about 4- 18 %. Acetic acid has a sour taste with a pungent smell just like most organic acids. This is the characteristic tart flavor of vinegar containing salad dressings. The acid has a molecular weight of 60.05, pKa value of 4.75, solidifies at -17°C, and boils at 118°C. It is miscible with alcohol, water, glycerol, ether, acetone, benzene, carbon tetrachloride, and insoluble in carbon disulfide [1]. The name acetic is derived from acetum, the Latin word for vinegar. Acetic acid is thought to have been produced accidentally as a byproduct of wine making, through the action of micro-organisms [2-4]. Naturally occurring bacterial species like *Acetobacter aceti* produce acetic acid through fermentation [5]. When fruits are fermented over a long period of time the alcohol produced is converted to acetic acid. At high concentrations acetic acid is corrosive and can cause severe burns when in contact with human skin. However, in its dilute form it can be used to preserve foods against spoilage bacteria. The global demand for acetic acid is around 6.5 million tonnes per year (Mt/a). About 5 Mt/a is manufactured from petrochemical feedstock and the remaining 1.5 Mt/a is produced through re-cycling. As already explained, vinegar is the dilute form of acetic acid, produced by fermentation and subsequent oxidation of ethanol to acetic acid and other short chain organic acids [3, 4, 6, 7]. However, this biological process of acetic acid fermentation is expensive in terms of industrial production, and so, much of the commercial acetic acid is produced through the Monsanto or Cativa processes (catalytic carbonylation of methanol).

History of Acetic Acid Production

In the pure form acetic acid is known as "glacial". This is because in the ancient times acetic acid manufacturers separated the acid from water by chilling the solution and as a result the acid solidified into crystals. For this reason, water free (anhydrous) acetic acid is known as glacial acetic acid. This is similar to the German name "*Eisessig*", which means ice-vinegar. The name is derived from the ice-like crystals that form below room temperature (about 16 °C). The trivial name, acetic acid is commonly used than ethanoic acid, the systematic (IUPAC) name. Around 721-815 B.C, an Arabian alchemist named Jabir Ibn Hayyan Geber produced strong acetic acid by distilling vinegar. However it wasn't until the 17th century that Georg Ernst Stahl isolated acetic acid in its most pure form. At about the same time, the 17th century scientists originally thought that acetic acid and other organic compounds could only be produced from compounds that occur in living organisms. This theory was, however, disproved in the middle of the 18th century following the synthesis of organic substances from chemicals. In particular, Hermann Kolbe a German organic chemist synthesized acetic acid from chemicals [8]. The Hermann Kolbe synthesis reaction sequence involved chlorination of carbon disulfide to carbon tetrachloride. Which was further converted to tetrachloroethylene through pyrolysis. The resultant tetrachloroethylene was chlorinated to trichloroacetic acid that was reduced to acetic acid through electrolysis.

A century later (18th century) Konrad Emil Bloch, a German-born American chemist, showed that acetic acid was the major precursor of cholesterol. Cholesterol is a lipid molecule abundantly found in animal cell tissues with functions in membrane permeability and fluidity as well as biosynthesis of sterols, bile acids and vitamin D [9]. Naturally, acetic acid is converted to cholesterol in the liver through a series of steps and Bloch was able to determine which of cholesterol's carbon atoms came from acetic acid. Through his research we are today able to understand the principles behind cholesterol metabolism [10, 11].

Chemistry of Acetic Acid

The functional carboxylic acid group of organic compounds is characterized by a carboxyl group which combines the carbonyl group of aldehydes and ketones with the hydroxyl group of alcohols and phenols. Since the carbonyl and the hydroxyl groups are directly bonded to each other, each affects the properties of the other. Most carboxylic acids are considered to be weak acids. Acetic acid is the simplest carboxylic acid after formic acid in which a single hydrogen atom is attached to the carboxyl group. When a methyl group is attached to the carboxyl group the compound becomes acetic acid. The acid is completely miscible with water, ethyl alcohol and ether, but insoluble in carbon disulfide. For example acetic acid ionizes to less than 1 % when dissolved in water. An aqueous solution of acetic acid like vinegar is a weak acid that is not completely ionized in the solvent [12]. The degree of ionization depends on the formal concentration of the acid as well as the existence of other acid or base species that are in the solution. Carboxylic acids react with alcohols in the presence of strong acids, which act as catalyst to form esters and water molecules in a process known as fisher esterification. Acetic acid reacts with isopentanol (Isoamyl alcohol) to form isopentyl acetate ester that is sometimes called banana oil, because the oil gives bananas their flavor and fragrance [13]. Many esters like isopentyl acetate have pleasant fragrances and are therefore used in perfumes and flavorings.

The general acetic acid formula is often written as $CH_3–C(O)OH$, $CH_3–C(=O)OH$, CH_3COOH or CH_3CO_2H, and abbreviated as AcOH, where Ac stands for the acetyl group $CH_3–C(=O)–$. The Acetate anion ($CH_3COO–$) is abbreviated as AcO–. In the context of acid-base reactions, however, the abbreviation HAc is used. Here Ac stands for acetate, the conjugate base. Acetate refers to a salt containing acetic acid anion, or ester. Acetic acid content in vinegar can be determined through an acid-base titration method to yield a salt and water. The reaction depends on the concentration (molarity) of the acid. Acetic acid ionizes fully to acetate at physiological pH. The acetate ion and acetic acid are at equilibrium, i.e. the pH of the acid equals its pKa. The acetyl group is essential in all biological systems. It is important in the metabolism of carbohydrates and lipids when bound to coenzyme A. Unlike other long chain carboxylic acids, acetic acid does not occur in natural triglycerides. However, the artificial triglyceride triacetate is commonly added to foods, cosmetics and topical medicines. Acetic acid is classified as a monoprotic acid, with one acidic hydrogen atom as shown in the Lewis structure of the acid and acetate ion (Figure 1).

(a)

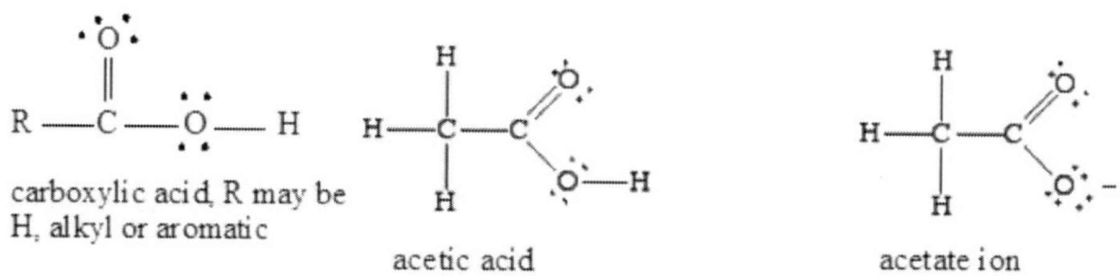

(b) $CH_3COOH + H_2O \rightleftharpoons H_3O^+ + CH_3COO^-$

(c) $$Ka = \frac{[CH_3COO^-][H_3O^+]}{[CH_3COOH]}$$

Figure 1. Equilibrium between acetic acid and acetate ion.

Table 1. Ka, and pKa values of Acetic Acid and its Derivatives

Acid	HA	A^-	Ka	pKa
Acetic acid	CH_3COOH	CH_3COO^-	1.76×10^{-5}	4.75
Chloroacetic acid	$CH_2ClCOOH$	CH_2ClCOO^-	1.41×10^{-3}	2.85
Dichloroacetic acid	$CHCl_2COOH$	$CHCl_2COO^-$	5.13×10^{-2}	1.29
Trichloroacetic acid	$CHCl_3COOH$	$CHCl_3COO^-$	2.24×10^{-1}	0.65
Triflouroacetic acid	CHF_3COOH	CHF_3COO^-	3.16×10^{-1}	0.5

When acetic acid is dissolved in water it establishes equilibrium between the acid and the acetate ion, in which Ka is the equilibrium constant (Figure 1).

Ka is the acetic acid dissociation constant, calculated to be 1.76×10^{-5}. The calculated pKa equals 4.75. The magnitude of the Ka and pKa values of weak acids indicates their relative strength. Chemically, weak acids exhibit less dissociation to the conjugate base and the equilibrium favors the undissociated weak acid form that yields a smaller Ka value. A smaller Ka usually corresponds to a larger pKa, given that pKa = – log Ka, which implies that weak acids have large pKa. The stoichiometric behavior of weak acids is beyond the scope of this chapter and further reading on Henderson-Hesselbalch relationship is recommended. The Henderson-Hasselbalch equation is extremely important in preparing buffer solutions of desirable pH values. The equation indicates that the pH of a buffer solution does not depend on the total concentration of the buffering acid and conjugate base but on the pKa and the ratio of the concentration of these two species. Thus, two structurally related molecules like acetic acid and chloroacetic acid have huge variations in their pKa values. This pKa reduction is observed with continued substitution of the hydrogen atom (Table 1).

Properties of Acetic Acid

Acetic acid is an organic acid in the carboxylic acid functional group, with the molecular formula CH_3-COOH.

It is a monocarboxylic acid because it contains only one –COOH functional group. Like all carboxylic acids it has a sour taste and a pungent smell. It is the main component of vinegar at 5-8 % of acetic acid solution. Vinegar is largely used as a food preservative and in the pickling of vegetables. Acetic acid dissociates in water to form hydrogen (H^+) ions. The hydrogen proton gives the acetic acid solution an acidic characteristic. Like all acidic solutions acetic acid turns blue litmus paper red, indicating that it is acidic in nature. However, this is a weak acid because it does not dissociate completely in aqueous solution. Acetic acid reacts with sodium bicarbonate to produce carbon dioxide (CO_2) gas that turns lime water milky due to formation of a solid calcium carbonate (Figure 2).

Production of Acetic Acid

A decade ago the total world acetic acid production was 5 Mt/a. The United States of America produced half the amount and the rest was shared between Europe and Japan. An annual recycling of about 1.5 Mt of acetic acid brought the total global production to 6.5 Mt/a. The global production has since increased to 10.7 Mt/a. Major global producers of acetic acid are Celanese and BP Chemicals, Millennium Chemicals, Sterling Chemicals, Samsung, Eastman and Svensk Etanolkemi. Industrial production of acetic acid is both synthetic and fermentative, whereby 75 % of the acid is by methanol carbonylation (Monsanto process). The bacterial fermentation process accounts for about 10 % of the total world production. Despite this low volume, the fermentation process remains the most important process in vinegar production. This is because food grade vinegar should be of biological origin according to food regulations and standards.

(a) $$CH_3COOH_{(aq)} + NaHCO_{3(aq)} \longrightarrow CH_3COONa_{(aq)} + CO_{2(g)} + H_2O_{(l)}$$

(b) $$\underset{\text{Lime water}}{Ca(OH)_{2(aq)}} + CO_{2(g)} \longrightarrow \underset{\text{Calcium carbonate}}{CaCO_{3(s)}} + H_2O_{(l)}$$

Figure 2. Reaction of acetic acid with sodium carbonate.

Oxidation of Acetaldehyde to Acetic Acid

Subsequent to the commercialization of the Monsanto and Cativa processes, acetic acid was produced by oxidation of acetaldehyde. Acetaldehyde oxidation is the second most important method of acetic acid production after methanol carbonylation. Organic compounds like butane or light naphtha are first oxidized at high temperature to acetaldehyde in the presence of air and metal ions like manganese, cobalt and chromium. The resultant acetaldehyde is oxidized to acetic acid.

Methanol Carbonylation to Acetic Acid

Methanol can be converted to acetic acid through a process known as carbonylation. Methanol carbonylation to acetic acid is catalysed with high rates at low water concentrations using an iridium/iodide based catalyst [14, 15]. Methanol reacts with carbon monoxide by the following equation to form acetic acid.

$$CH_3OH + CO \longrightarrow CH_3COOH$$

Although various catalysts have traditionally been used in both the BASF and Monsanto processes of methanol carbonylation, a new process concept has been employed by BP Chemicals (Formally Monsanto). Rhodium is used to catalyze the carbonylation of methanol/ methyl acetate mixtures to yield acetic acid and acetic anhydride. In the industrial process, methanol and CO react continuously under controlled pressure (3000 KPa) and temperature (150-200°C) to yield acetic acid and CO_2 and H_2 as the main byproducts. The Monsanto process has, however, been succeeded by the Cativa process in the production of acetic acid. The catalyst of choice in the Cativa process is a combination of Iridium/ruthenium as opposed to rhodium use in the Monsanto process.

The Iridium/ruthenium complex is superior to the rhodium based systems because it allows use of less water in the reaction mixture. The quality of acetic acid produced under low water concentration is exceptionally good. It has also the advantage of a reduced number of drying columns as well as byproducts like propionic acid and it suppress the water-gas shift reaction [14, 15].

Acetic Anhydride

About 30 % of the global acetic acid is used in the production of acetic anhydride. Acetic anhydride is formed when two molecules of acetic acid condense. This is a substitution reaction whereby there is loss of water from two acetic acid molecules. Normally acetic acid is dehydrated to form a ketene, which condenses in the presence of acetic acid to form the acetic anhydride. Acetic anhydride can also be formed by the reaction of an acid chloride and a salt of acetic acid like sodium acetate (Figure 3). Acetic acid is also produced by carbonylation of methyl acetate in a process similar to the Monsanto process but under anhydrous conditions [16]. Given that acetic anhydride is mainly used in the acetylation process, it is a good agent in the production of cellulose acetate [17], a synthetic textile, which is also useful for photographic film.

Acetic anhydride is also used as a reagent in production of heroin and other compounds as well. Cellulose, which is a component of cotton and wood, is rich in hydroxyl groups that readily react with acetic anhydride to form cellulose acetate. The resultant cellulose acetate is used as a base for magnet films and manufacture of textile fibers. Examples include photographic films and rayon which are made from cellulose acetate fibers. When acetic acid is heated in the presence of salicylic acid it forms acetylsalicylic acid (aspirin). This is a reversible esterification process; however, presence of water (Figure 4) can lead to hydrolysis of aspirin. Thus use of acetic anhydride gives better yields due to absence of water.

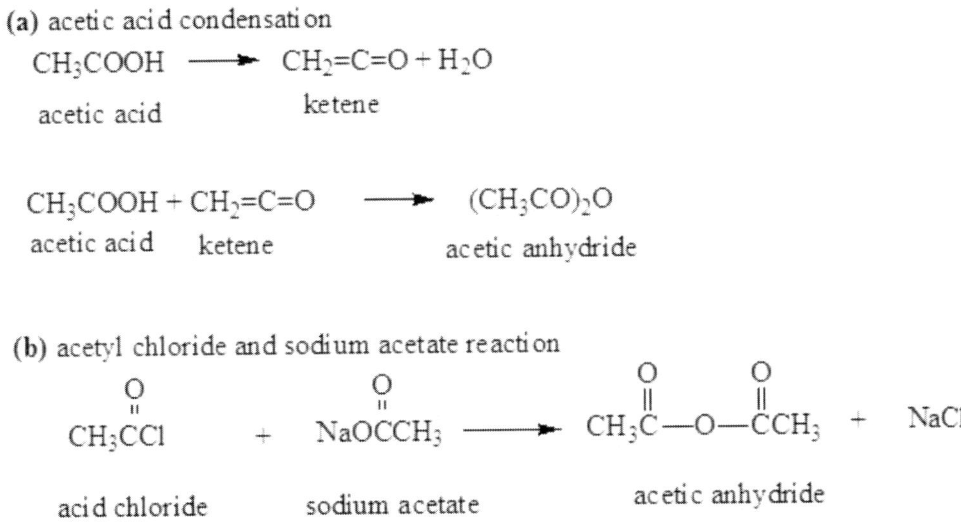

Figure 3. Formation of acetic anhydride.

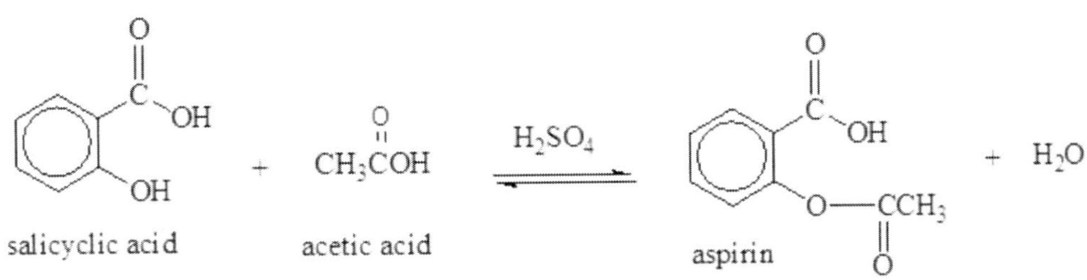

Figure 4. Reaction of salicylic acid and acetic acid to form aspirin.

Vinyl Acetate

Acetic acid is also largely used as a raw material in the production of vinyl acetate monomer (VAM). Production of VAM consumes about 45 % of the global acetic acid produced. Acetic acid reacts with ethylene and oxygen or with acetylene in the presence of a palladium catalyst to form vinyl acetate that polymerizes to polyvinyl acetate (Figure 5) by itself and with other monomers to form copolymers such as ethylene-vinyl acetate (EVA), vinyl acetate-acrylic acid (VA/AA) and polyvinyl chloride acetate (PVCA). The resultant acetate polymers are important resins used in paints, adhesives, plastics and textile finishes. Other uses of acetic acid include the manufacture of pigments, dyes and pesticides. Acetic acid is used to manufacture esters of ethyl and butyl acetate (Table 2).

Acetate esters are good solvents for many resins. Reaction of acetic acid and chlorine produces chloroacetic acid; which in turn reacts with alkali cellulose to form carboxymethylcellulose. Chloroacetic acid is also used in the manufacture of phenoxy herbicides such as 2,4-dichlorophenoxyacetic acid and 2,4,5-trichlorophenoxyacetic acid.

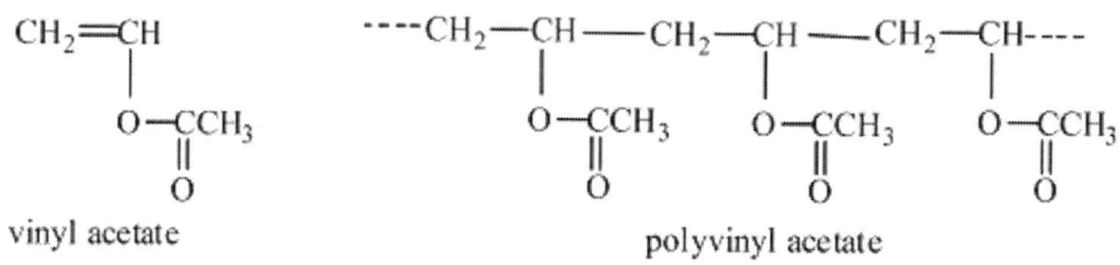

Figure 5. Polymerization of vinyl acetate.

Table 2. A summary of Physical and Chemical Properties of Acetic Acid and its Derivatives

Chemical	Formula	M.W	B.P (°C)	M.P (°C)	Density	Solubility (g/100g)
Acetic Acid	$C_2H_4O_2$	60	118	16.6	1.05	Miscible
Acetic Anhydride	$C_4H_6O_3$	102	139	-73.1	1.08	12
Acetone	C_3H_6O	58	56.2	-94.3	0.79	Miscible
Acetonitrile	C_2H_3N	41	81.6	-46	0.79	Miscible
Butyl Acetate	$C_6H_{12}O_2$	116	125	-106	0.88	0.68
Ethyl Acetate	$C_4H_8O_2$	88	77	-83.6	0.90	8.7

M.W: molecular weight, M.P: melting point, B.P: boiling point.

Sodium Acetate

Acetic acid reacts with sodium carbonate or sodium hydroxide to form sodium acetate (sodium ethanoate). Sodium acetate is used as a buffer in petroleum production, food preservations, and electroplating. It is used in the tanning industry to enhance penetration of the tan in the leather and as a mordant in the dyeing processes. Sodium acetate is also used as an intermediate in the manufacture of dyes and pharmaceutical products [18], in glucose purification and soap production. Sodium acetate is also gaining use in treatment of poisoning as a replacement to sodium bicarbonate in medical toxicology [19].

Acetate Ester

Production of acetate esters is based on the esterification process of reacting a carboxylic acid and an alcohol. Examples include ethyl acetate, butyl acetate, and propyl acetate. These esters are used as industrial solvents in the manufacture of inks and paints. For instance ether acetates are used as solvents for nitrocellulose, acrylic lacquers, varnish removers and wood stains.

Most acetate esters, however, are produced from acetaldehyde using the Tishchenko reaction. The reaction of ethylene oxide or propylene oxide with alcohol produces glycol ethers. These are then esterified with acetic acid to form ethylene glycol monoethyl ether acetate (EEA), ethylene glycol monobutyl ether acetate (EBA), and propylene glycol monomethyl ether acetate (PMA, a resist solvent).

Chemical Reactions

Acetic acid exhibits the behavior of carboxylic acids and it reacts with a standard base to form metal acetates and water. However, reactions with strong bases like organolithium somewhat yields organometallic intermediates that are reduced to ethanol (Figure 6). The hydroxyl (OH) functional group is the main reaction site as illustrated by the conversion of acetic acid to acetyl chloride (Figure 6). Esters of acetic acid can likewise be formed via Fischer esterification, and amides can be formed. When heated above 440 °C, acetic acid decomposes to produce carbon dioxide and methane, or to produce ketene and water.

Commercial Vinegars

Vinegar is a dilute form of acetic acid, about 4-18 % acetic acid (w/w) produced by either fast or slow fermentation of ethanol by acetic acid bacteria [4, 7]. The fermentation process can take weeks or even months depending on various factors and fermentation conditions, which can be manipulated to produce the desirable product. Vinegars have traditionally been fermented from a variety of raw materials including foods such as fruits, vegetables, cereals and root crops [20]. The chemical process involves partial oxidation of ethanol to acetaldehyde that is eventually converted to acetic acid.

Beer, Malt and Distilled Vinegar

Unlike balsamic vinegar with Italian roots, beer vinegar is mainly produced in the United Kingdom, German, Austria and the Netherlands. Although the beer vinegar is said to have a malty taste, its flavor is dependent on the type of beer used. In contrast to beer, malt vinegar or 'Alegar' is produced by malting barley.

The starch in the malted barley is converted to maltose, which is brewed to ale beer that is converted to vinegar that is allowed to age to develop taste and flavor [21]. "Distilled vinegar" is obtained by fermenting distilled alcohol and not distillation of vinegar as the name may suggest. The distilled vinegar is a colorless solution of about 5-8 % acetic acid with a pH of about 2.4 and it is commonly referred to as distilled spirit or white vinegar. It is has various uses including medicinal, laboratory, cleaning, cooking, baking, food preservation and pickling.

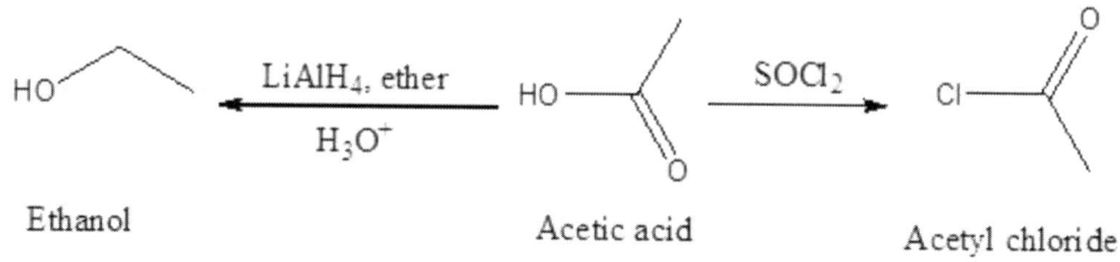

Figure 6. Reaction of acetic acid.

Cane, Coconut and Palm Vinegars

Cane vinegar is made from sugarcane juice [22] and spirit vinegar is a strong variety of cane vinegar (5-20 % acetic acid by mass). It is has a dark yellow to golden brown color, with a mellow flavor similar to rice vinegar. Coconut water is fermented to make coconut vinegar. It has a cloudy white color with a sharp acidic taste with a tinge of yeast. *Sinamak* vinegar is a variation of cane or coconut vinegar in the Philippines. To make *Sinamak* cane and coconut vinegars are mixed with spices. Palm vinegar is produced from fermented sap from flower clusters of nipa palm or attap palm in the Philippines with a mild pH of about 5.5.

Fruit Vinegars

Most fruit vinegars are made from fruit wines in which the original fruit flavor is maintained. Examples include apple, raspberry, tomato, date, and kiwifruit. Europe produces most fruit vinegars because of increasing market demand as opposed to non-fruit vinegars that are infused with fruits or fruit flavor. Persimmon vinegar is made in South Korea while *Jujube* and wolfberry vinegars are made in China. Apple cider vinegar is produced from apple must or cider. It is brown in color with a pH of about 4.5 that makes it mildly acidic. The acidity makes it very harsh with a burning sensation to the throat and thus it may be diluted with honey to mask the acidity. *Jamun sirka* vinegar is produced from rose apple in India and it is considered to have medicinal effects in gastrointestinal, spleen and diabetic ailments. Raisin or grape vinegar is produced from raisins in the Middle East. It is cloudy and medium brown in color with a mild flavor. Vinegar can be flavored or infused with fruit to make fruit flavored vinegar or spiced with herbs like oregano, chilies or thyme to make herb vinegar. Balsamic vinegar is produced from the concentrated juice or must of white grapes. It is dark brown with a rich, sweet and complex tart flavor with high acidity levels[23]. Originally an Italian product, it was fermented for years in successive casks made of various types of wood like oak, chestnut, ash and acacia. Regardless of how it is produced, balsamic must be made from grapes.

Rice and Chinese Black Vinegar

Rice vinegar is available in different color varieties. For instance fermentation of unpolished brown rice produces brown rice vinegar. Light rice vinegar is popular in Japan for preparation of sushi rice and salad dressings probably due to its mild acidity and flat flavor. Black rice vinegar that is made from black glutinous rice is popular in China and other East Asian countries. Some varieties of rice vinegar are sweetened or otherwise seasoned with spices or other added flavorings. Job's tears vinegar is made from Job's tears; a tall, grain-bearing, tropical plant in Japan, its flavor is similar to that of rice vinegar. "East Asian black" or Chinese black vinegar is made from cereals like rice, wheat, millet, sorghum or a combination of both. It has an inky characteristic black color with a malty flavor. Examples include *Zhenjiang* and *Shanxi* vinegars from China and *Kurosu* from Japan. Since 2004, *Kurosu* has been marketed as a health promoting product containing high concentrations of amino acids and recent research shows that it has both *in vivo* and *in vitro* anticancer and anti-colitis properties [24, 25]. In particular, *Kurosu* vinegar is rich in phenolic compounds, and the in-vitro antioxidant activity of an ethyl acetate extract of *Kurosu* vinegar suppressed lipid peroxidation in a similar manner to alpha-tocopherol [26].

Wine Vinegar

This is made from both red and white wines and just like the wines there is a variation in quality. Better quality wine vinegars are matured in wood barrels for up to two years and exhibit complex and mellow flavor. Examples include champagne, pinot grigio and sherry [27]. Sherry vinegar has a dark mahogany color with a stronger flavor. Sherry vinegar is used in vinaigrettes and salad dressings as a flavoring.

Table 3. Summary of Selected Commercial Vinegars

Vinegar	Source	Appearance	Flavor	pH	Origin
Apple cider	Cider	Golden brown	Sweet and sour	4.5	United Kingdom
Balsamic	Grape juice	Brown/black	Sweet and sour	< 3	Italian
Beer	Beer	Light golden	Sharp and malty	N.D	United Kingdom, Germany, Austria
Cane	Sugarcane	Dark yellow/ golden brown	Mellow	N.D	Philippines
Chinese black	Rice, sorghum	Inky black	Malty	N.D	China
Coconut	Coconut water	Cloudy white	Sharp, yeasty	N.D	Philippines
Distilled	Distilled alcohol	Colorless	Sharp and acidic	2.4	US
Malt	Malted barley	Light brown	Malty	N.D	United Kingdom
Rice	Rice	White, red, black	Mild acidity	N.D	Asian
Palm	Palm flower sap	Light brown	Mild acidity	<5.5	Asia, Africa
Wine	Red/ white wine	Brown	Mellow	N.D	Mediterranean, Central Europe

N.D.= Not Determined.

Kombucha Vinegar

This is made from *kombucha*, a mixed culture of yeast and bacteria [28]. To develop flavor strawberries, blackberries or blueberries are added at the beginning of the fermentation process. Although no scientific studies have been done to prove its health benefits, it is said that *kombucha* vinegar promotes a healthy digestive tract due to the diverse microbial composition.

Acetic Acid Applications

Acetic acid is one of the simplest carboxylic acids. It is an important industrial chemical, mainly used in the production of cellulose acetate [29] for photographic film and polyvinyl acetate for wood glue, as well as synthetic fibers and fabrics [30]. In the food industry, acetic acid is used under the food additive code E260 as an acidity regulator and as a condiment [1]. Acetic acid is used in the formulation of hair conditioners, shampoos and other skin care products as well as mouthwashes [31]. The dilute form of acetic acid is used as a descaling agent in household cleaning. Acetic acid's importance is not limited to its role in biological processes. White or distilled vinegar is used as a household cleaner and in food processing as a preservative [32-34]. Acetic acid is an important commercial chemical as well since it can be used as a solvent or converted into such useful compounds as acetic anhydride and acetate esters used in the manufacture of rayon fiber. Acetic acid is also important in the manufacture of synthetic fabrics [29], pharmaceuticals [18], dyes and perfumes. It is also used as a coagulant in the manufacture of rubber. It has also been used historically as a fungicide in certain agricultural application. Acetic Acid is formed when ethanol is fermented and is referred to as vinegar due to the low acid concentration. Many of the naturally fermented vinegars from various raw materials also contain small amounts of organic acids such as lactic, tartaric and citric acids [35].

Acetic Acid As a Solvent

Glacial acetic acid is an excellent polar protic solvent that is frequently used to purify organic compounds by re-crystallization. It is also used as a solvent in the production of terephthalic acid [36], the raw material for polyethylene terephthalate. Other uses involve carbocation reactions– Friedel-Crafts alkylation– as a solvent. In the production of synthetic camphor for instance, acetic acid acts both as a solvent and as a nucleophile to trap the rearranged carbocation. It is also used as a solvent in the reduction of aryl nitro-group to aniline using palladium-on-carbon. In analytical chemistry acetic acid is used to estimate weak alkaline substances like amides. Acetic acid is a weak base compared to water hence the amide behaves as a strong acid such as perchloric acid.

Summary of Special Applications and Uses of Acetic Acid

- Component of stop solution used in development of photographic film.
- A de-scaling agent used to remove lime-scale from taps and kettles.

- Lyses of red blood cells during smear preparation for microscopic examination.
- Cleaning of indoor climbing holds made of chalk (magnesium carbonate).
- Treatment of box jellyfish stings. The acidity disables the stinging cells of the jellyfish, preventing serious injury or death if applied immediately.
- Preservation of livestock silage against spoilage bacteria.
- Fumigant in bee hives for pest control.
- Removal of wart and verruca.
- In physical therapy to break up nodules of scar tissue through iontophoresis.

Table 4. Applications of Inorganic and Organic Salts of Acetic Acid

Salt	Application
Sodium acetate	Textile and food (as a preservative) industries
Copper (II) acetate	Color pigment and fungicide
Aluminium acetate	Mordant for dyes
Iron (II) acetate	Mordant for dyes
Palladium (II) acetate	Catalyst for organic coupling reactions e.g. Heck reaction
Silver acetate	pesticide
Monochloroacetic acid	Manufacture of indigo dye
Bromoacetic acid	When esterified it forms ethylbromoacetate
Trifluoroacetic acid	A reagent in organic synthesis

Acetic Acid in Meat Preservation

The development of chemical methods of food preservation has become important. The number of antimicrobial chemical preservatives approved for use in foods is limited due to low solubility and the negative effect on food sensory properties. Acetic, propionic, sorbic and citric acids are used broadly as antimicrobial agents, because of their solubility and bacteriostatic or bacteriocidal properties [33, 34]. Among the chemicals used in long-term preservation of food, acetic acid should be considered because of its unique advantages. It is widely available and cost-effective compared to other acidulants [32]. Acetic acid is listed among the generally recognized as safe substances by the US Food and Drug Administration (FDA). That is as a multipurpose food additive, a substance migrating to food from cotton and cotton fabrics used in dry-food packaging, a substance migrating to food from paper and paperboard products, and a general purpose food additive for animal feed. In a marinated chicken experiment, vinegar reduced populations of *Campylobacter jejuni* while maintaining an acceptable taste and texture [37]. The partial dissociation of acetic acid plays an important role in its ability to inhibit microbial growth. The action of the acid is based on lowering the pH value of the product to be preserved. Above a concentration of 0.5%, the acid displays antimicrobial action in penetrating the cell wall and denaturing the protein of the cell plasma. This affects the cell where essential cell functions such as ATP synthesis in bacteria, active transport of nutrients and cytoplasmic regulation take place [32].

Medicinal Effects of Acetic Acid

Although data on evidence-based research is still scanty some studies demonstrate that acetic acid has some merit as a therapeutic compound when used judiciously. In one study, vinegar was shown to reduce postprandial glycemia in healthy adults. The antiglycemic properties of vinegar were evident when small amounts of vinegar were ingested with meals composed of complex carbohydrates [38, 39]. Vinegar supplementation was shown to lower glucose and insulin responses while increasing satiety. However, it is still unclear how vinegar alters meal-induced glycemia. In another study, detection and recognition threshold of salt were decreased in salt solutions with added the vinegar ingredient.

This tendency was more pronounced with rice black vinegar than with rice vinegar. However, no significant changes in the threshold of both detection and recognition were observed when salt at the half concentration of the detection threshold was added to rice vinegar, indicating that vinegar could be used to reduce salt intake in the diet [40]. Acetates are common constituents of animal and plant tissues formed following food metabolism. Acetic acid is absorbed from the gastrointestinal tract and through the lungs. It is readily metabolized by most tissues and may give rise to ketones as intermediates. *In vitro* studies have shown that acetate is incorporated into phospholipids, sterols, and fatty acids in a variety of animal tissue preparations.

Animal studies show that a diet containing acetic acid or vinegar at concentrations found in traditional diets has such effects as enhancement of glycogen repletion [41, 42], prevention of hypertension [43], cancer [44] and stimulation of Calcium absorption [45]. In one *in vivo* study where rats were fed a diet containing cholesterol, acetic acid reduced serum total cholesterol and triacylglycerol. This was due to inhibition of lipogenesis in the liver and increased excretion of fecal bile acid [46]. Taken together, the scientific evidence linking vinegar use to reduced risk for cancer, hypertension and diabetes is unclear and more investigations in carefully controlled clinical trials are needed to describe the mechanisms of action.

Health and Safety Considerations

Concentrated acetic acid is corrosive and should be handled with care to avoid causing skin burns, eye damage, and/or irritation to the mucous membranes. These burns appear long after exposure, so special resistant gloves such as nitrile rubber should be worn. Use of latex gloves should be discouraged as they offer no protection when handling the compound. The hazards of solutions of acetic acid depend on the concentration (Table 5).

Whereas dilute acetic acid (vinegar) is harmless, solutions above 25% by mass acetic acid should be handled in a fume hood because of the pungent, corrosive fumes. Due to incompatibility, it is recommended to keep acetic acid away from chromic acid, ethylene glycol, nitric acid, perchloric acid, permanganates, peroxides and hydroxyls. Based on animal experiments and industrial exposure acetic acid is a strong eye, skin, and mucous membrane irritant, and Ingestion of concentrated solutions can cause severe damage to the gastrointestinal tract as well as adversely affect the blood pH.

The Occupational Safety and Health Administration (OSHA) standard for airborne acetic acid as an eight-hour, time-weighted average (TWA) is 10 ppm. However, acetic acid has not been shown to be carcinogenic, teratogenic or mutagenic in animal studies.

Table 5. European Union Classification of Acetic Acid Solutions

Concentration (w/w)	Molarity	Classification	R-Phrases
10–25%	1.67–4.16 mol/L	Irritant (Xi)	R36/38: Irritating to eyes and skin
25–90%	4.16–14.99 mol/L	Corrosive (C)	R34: Causes burns
>90%	>14.99 mol/L	Corrosive (C) Flammable (F)	R35: Causes severe burns R10: Flammable

On the contrary Sodium fluoroacetate (CF_3COONa) and fluoroacetamide (CF_3CONH_2) are used mainly as rodenticides display some level of toxicity. Fluoroacetic acid is the toxic principle of Dichapetalum cymosum (gifblaar), a plant endemic to southern Africa [47]. Fluoroacetates block the tricarboxylic acid cycle and are extremely toxic [48, 49].

Conclusion

Acetic acid is a simple carboxylic acid, in which a methyl group is attached to the carboxyl group. Industrial production of acetic acid is largely through the Monsanto process. It is an important industrial solvent and can be converted into other compounds such as acetic anhydride and acetate esters that are useful in the manufacture of cellulose acetate, rayon, and sodium acetate. Vinegar is produced by acetic acid bacteria that ferment ethanol and is largely used as a food preservative and in the pickling of vegetables. Even though the scientific data is still scanty, acetic acid has been shown to alleviate pain in a variety of diseases including cancer, diabetes and hypertension, but more scientific studies are needed to delineate the mechanisms of action. Besides the many beneficial effects acetic acid should be handled with care because of health and safety considerations; concentrated acetic acid is corrosive and can cause skin burns or irritation to the mucous membranes.

References

[1] JECFA (1997) Compendium of food additive specifications, forty-ninth report of the joint FAO/WHO expert committee on food additives. *FAO food and nutrition* paper 5, 212-213.

[2] Benziman, M., Abeliovitz, A. (1964) Metabolism of Dicarboxylic Acids in Acetobacter Xylinum. *Journal of Bacteriology* 87, 270-7.

[3] Ajl, S. J. (1951) Studies on the mechanism of acetate oxidation by bacteria. V. evidence for the participation of fumarate, malate, and oxalacetate in the oxidation of acetic acid by Escherichia coli. *Journal of General Physiology* 34, 785-94.

[4] Nakayama, T. (1961) Studies on acetic acid bacteria. IV. Purification and properties of a new type of alcohol dehydrogenase, alcohol-cytochrome-553 reductase. *Journal of Biochemistry* 49, 240-51.

[5] Parmar, I., Rupasinghe, H. P. (2013) Bio-conversion of apple pomace into ethanol and acetic acid: Enzymatic hydrolysis and fermentation. *Bioresource Technology* 130, 613-20.

[6] Leisinger, T. (1965) [Studies on the systematics and metabolism of acetic acid bacteria]. *Zentralbl. Bakteriol. Parasitenkd Infektionskr. Hyg.* 119, 329-76.

[7] Nakayama, T. (1961) Studies on acetic acid bacteria. III. Purification and properties of coenzyme-independent aldehyde dehydrogenase. *Journal of Biochemistry* 49, 158-63.

[8] Rocke, A. J. (1993) *The Quiet Revolution: Hermann Kolbe and the Science of Organic Chemistry*. University of California Press, Berkeley.

[9] Hanukoglu, I. (1992) Steroidogenic enzymes: Structure, function, and role in regulation of steroid hormone biosynthesis. *The Journal of Steroid Biochemistry and Molecular Biology* 43, 779-804.

[10] Dietschy, J. M. (1984) Regulation of cholesterol metabolism in man and in other species. *Klin. Wochenschr.* 62, 338-45.

[11] Dietschy, J. M., Wilson, J. D. (1970) Regulation of cholesterol metabolism. 3. *N Engl. J. Med.* 282, 1241-9.

[12] Partanen, J. I., Covington, A. K. (2003) Determination of stoichiometric dissociation constants of acetic acid in aqueous solutions containing acetic acid, sodium acetate, and sodium chloride at (0 to 60) C. *Journal of Chemical Engineering Data* 48, 797-807.

[13] Yilmaztekin, M., Erten, H., Cabaroglu, T. (2009) Enhanced production of isoamyl acetate from beet molasses with addition of fusel oil by Williopsis saturnus var. saturnus. *Food Chemistry* 112, 290-294.

[14] Jones, J. H. (2000) The Cativa™ process for the manufacture of acetic acid. *Platinum Metals Review* 44, 94-105.

[15] Sunley, G. J., Watson, D. J. (2000) High productivity methanol carbonylation catalysis using iridium: The Cativa™ process for the manufacture of acetic acid. *Catalysis Today* 58, 293-307.

[16] Zoeller, J. R., Agreda, V. H., Cook, S. L., Lafferty, N. L., Polichnowski, S. W., Pond, D. M. (1992) Eastman chemical company acetic anhydride process. *Catalysis Today* 13, 73-91.

[17] Biswas, A., Saha, B. C., Lawton, J. W., Shogren, R. L., Willett, J. L. (2006) Process for obtaining cellulose acetate from agricultural by-products. *Cabohydrate Polymers* 64, 134-137.

[18] Fouad, E. A., El-Badry, M., Alanazi, F. K., Arafah, M. M., Al-Ashban, R., Alsarra, I. A. (2010) Preparation and investigation of acetyl salicylic acid-caffeine complex for rectal administration. *Pharmaceutical Development and Technology* 15, 249-57.

[19] Neavyn, M. J., Boyer, E. W., Bird, S. B., Babu, K. M. (2013) Sodium Acetate as a Replacement for Sodium Bicarbonate in Medical Toxicology: a Review. *Journal of Medical Toxicology*.

[20] Natera, R., Castro, R., de Valme Garcia-Moreno, M., Hernandez, M. J., Garcia-Barroso, C. (2003) Chemometric studies of vinegars from different raw materials and processes of production. *J. Agric. Food Chem.* 51, 3345-51.

[21] Jones, D. D., Greenshields, R. N. (1969) Volatile constituents of vinegar. I. a survey of some commercially available malt vinegars. *Journal of the Institute of Brewing* 75, 457-463.

[22] Kocher, G. S., Kaira, K. L., Phutela, R. P. (2006) Comparative production of sugarcane vinegar by different immobilization techniques. *Journal of the Institute of Brewing* 112, 264-266.

[23] Antonelli, A., Chinnici, F., Masino, F. (2004) Heat-induced chemical modification of grape must as related to its concentration during the production of traditional balsamic vinegar: a preliminary approach. *Food Chemistry* 88, 63-68.

[24] Shizuma, T., Ishiwata, K., Nagano, M., Mori, H., Fukuyama, N. (2011) Protective effects of Kurozu and Kurozu Moromimatsu on dextran sulfate sodium-induced experimental colitis. *Digestive Diseases and Sciences* 56, 1387-92.

[25] Fukuyama, N., Jujo, S., Ito, I., Shizuma, T., Myojin, K., Ishiwata, K., Nagano, M., Nakazawa, H., Mori, H. (2007) Kurozu moromimatsu inhibits tumor growth of Lovo cells in a mouse model in vivo. *Nutrition* 23, 81-6.

[26] Nishidai, S., Nakamura, Y., Torikai, K., Yamamoto, M., Ishihara, N., Mori, H., Ohigashi, H. (2000) Kurosu, a traditional vinegar produced from unpolished rice, suppresses lipid peroxidation in vitro and in mouse skin. *Bioscience Biotechnology and Biochemistry* 64, 1909-14.

[27] Tesfaye, W., Morales, M. L., Garcia -Parrila, M. C., Troncoso, A. M. (2002) Wine vinegar: technology, authenticity and quality evaluation. *Trends in Food Science and Technology* 13, 12-21.

[28] Yang, Z., Zhou, F., Ji, B., Li, B., Luo, Y., Yang, L., Li, T. (2010) Symbiosis between microorganisms from kombucha and kefir: Potential significance to the enhancement of kombucha function. *Applied Biochemistry and Biotechnology* 160, 446-55.

[29] Fan, G., Wang, M., Liao, C., Fang, T., Li, J., Zhou, R. (2013) Isolation of cellulose from rice straw and its conversion into cellulose acetate catalyzed by phosphotungstic acid. *Carbohydr. Polym.* 94, 71-6.

[30] Wang, L., Hong, Y., Li, J. X. (2012) Durability of running shoes with ethylene vinyl acetate or polyurethane midsoles. *J. Sports Sci.* 30, 1787-92.

[31] Mumcuoglu, K. Y. (1999) Prevention and treatment of head lice in children. *Paediatr. Drugs* 1, 211-8.

[32] Adams, M. R., Moss, M. O. (1995) *Food Microbiology*. The royal society of Chemistry, Trobridge, Wiltshire.

[33] Podolak, R. K., Zayas, J. F., Kastner, C. L., Fung, Y. C. (1996) Inhibition of Listeria monocytogenes and Escherichia coli 0157:H7 on beef by application of organic acids. *Journal of Food Protection* 59, 370-373.

[34] Siragusa, G. R., Dickson, J. S. (1992) Inhibition of Listeria monocytogenes on beef tissue by application of organic acids immobilised in calcium alginate gel. *Journal of Food Science* 57, 293-296.

[35] Wang, C., Han, H., Gu, X., Yu, Z., Nishino, N. (2013) A survey of fermentation products and bacterial communities in corn silage produced in a bunker silo in China. *Animal Science Journal*.

[36] Hashizume, H., Izumisawa, Y. (1986) Process for preparing terephthalic acid by the catalytic oxidation with air of p-xylene in a water containing acetic acid solvent. Volume 4593122 Mitsubishi Chemical Industries Limited, US.

[37] Birk, T., Gronlund, A. C., Christensen, B. B., Knochel, S., Lohse, K., Rosenquist, H. (2010) Effect of organic acids and marination ingredients on the survival of Campylobacter jejuni on meat. *Journal of Food Protection* 73, 258-65.

[38] Ostman, E., Granfeldt, Y., Persson, L., Bjorck, I. (2005) Vinegar supplementation lowers glucose and insulin responses and increases satiety after a bread meal in healthy subjects. *European Journal of Clinical Nutrition* 59, 983-8.

[39] Johnston, C. S., Steplewska, I., Long, C. A., Harris, L. N., Ryals, R. H. (2010) Examination of the antiglycemic properties of vinegar in healthy adults. *Annals of Nutrition and Metabolism* 56, 74-9.

[40] Hatae, K., Takeutchi, F., Sakamoto, M., Ogasawara, Y., Akano, H. (2009) Saltiness and acidity: detection and recognition thresholds and their interaction near the threshold. *Journal of Food Science* 74, S147-53.

[41] Fushimi, T., Tayama, K., Fukaya, M., Kitakoshi, K., Nakai, N., Tsukamoto, Y., Sato, Y. (2001) Acetic acid feeding enhances glycogen repletion in liver and skeletal muscle of rats. *Journal of Nutrition* 131, 1973-7.

[42] Fushimi, T., Tayama, K., Fukaya, M., Kitakoshi, K., Nakai, N., Tsukamoto, Y., Sato, Y. (2002) The efficacy of acetic acid for glycogen repletion in rat skeletal muscle after exercise. *International Journal of Sports Medicine* 23, 218-22.

[43] Kondo, S., Tayama, K., Tsukamoto, Y., Ikeda, K., Yamori, Y. (2001) Antihypertensive effects of acetic acid and vinegar on spontaneously hypertensive rats. *Bioscience, Biotechnology and Biochemistry* 65, 2690-4.

[44] Nanda, K., Miyoshi, N., Nakamura, Y., Shimoji, Y., Tamura, Y., Nishikawa, Y., Uenakai, K., Kohno, H., Tanaka, T. (2004) Extract of vinegar "Kurosu" from unpolished rice inhibits the proliferation of human cancer cells. *J. Exp. Clin. Cancer Res.* 23, 69-75.

[45] Kishi, M., Fukaya, M., Tsukamoto, Y., Nagasawa, T., Takehana, K., Nishizawa, N. (1999) Enhancing effect of dietary vinegar on the intestinal absorption of calcium in ovariectomized rats. *Bioscience, Biotechnology and Biochemistry* 63, 905-10.

[46] Fushimi, T., Suruga, K., Oshima, Y., Fukiharu, M., Tsukamoto, Y., Goda, T. (2006) Dietary acetic acid reduces serum cholesterol and triacylglycerols in rats fed a cholesterol-rich diet. *British Journal of Nutrition* 95, 916-24.

[47] Cooper, R. G. (2007) Poisoning in ostriches following ingestion of toxic plants--field observations. *Tropical Animal Health Production* 39, 439-42.

[48] Proudfoot, A. T., Bradberry, S. M., Vale, J. A. (2006) Sodium fluoroacetate poisoning. *Toxicology Reviews* 25, 213-9.

[49] Giannitti, F., Anderson, M., Caspe, S. G., Mete, A., East, N. E., Mostrom, M., Poppenga, R. (2013) An Outbreak of Sodium Fluoroacetate (1080) Intoxication in Selenium- and Copper-Deficient Sheep in California. *Veterinery Pathology*.

In: Acetic Acids
Editor: Angelo Basile

ISBN: 978-1-62948-217-0

Chapter 2

Acetic Acids: Indole-3-Acetic Acid (IAA), Indole-3-Butyric Acid (IBA), A-Naphthaleneacetic Acid (A-NAA) and 2,4 Dichlorophenoxy Acetic Acid (2,4-D): Chemical Properties, Production and Applications in Tissue Culture, on Rooting and on Biochemical Parameters

Virginia Sarropoulou,[*] Kortessa Dimassi-Theriou and Ioannis Therios
Laboratory of Pomology, Department of Horticulture, School of Agriculture, Aristotle University of Thessaloniki, Thessaloniki, Greece

Abstract

In this chapter, the effects of the auxins IAA, IBA, a-NAA and 2,4-D from the category of the acetic acids in tissue culture, on rooting, on photosynthetic pigments (total leaf chlorophyll (a+b) and carotenoids content) as well as on total carbohydrate and proline concentration were analysed. Acetic acid is a weak organic compound with the chemical formula CH_3CO_2H (also written as CH_3COOH or $C_2H_4O_2$) and is the main component of vinegar. Acetic acid is mildly corrosive to metals including iron, magnesium, and zinc, forming hydrogen gas and salts called acetates. At physiological pHs, acetic acid is usually fully ionized to acetate. In biochemistry, acetate and acetic acid are equivalent. Acetic acid is produced industrially both synthetically and by bacterial fermentation. Glacial acetic acid is an excellent polar protic solvent, is often used as a solvent for reactions involving carbocations, such as Friedel-Crafts alkylation

[*] Corresponding author: Virginia Sarropoulou. E-mail: vsarrop@gmail.com; vsarropo@agro.auth.gr.

and is used in analytical chemistry for the estimation of weakly alkaline substances such as organic amides. Auxins derive their name from the Greek word *auxein* - "to grow/increase".

The most important member of the auxin family is indole-3-acetic acid (IAA). Synthetic auxin analogs include 1-naphthaleneacetic acid, 2,4-dichlorophenoxyacetic acid (2,4-D), and many others. Auxins are also often used to promote initiation of adventitious roots, to promote uniform flowering and fruit set, and to prevent premature fruit drop. On the cellular level, auxin is essential for cell growth, affecting both cell division and cellular expansion. Auxin participates in phototropism, geotropism, hydrotropism and other developmental changes. Auxins promote root initiation, induce shoot apical dominance and delay fruit senescence. In low concentrations, auxins can inhibit ethylene formation and transport in plants; however, high concentrations can induce the synthesis of ethylene. IAA is chemically similar to the amino acid tryptophan which is generally accepted to be the molecule from which IAA is derived. 2,4-D is absorbed through the leaves and is translocated to the meristems of the plant. Auxins and their synthetic analogues are commonly used for rooting of cuttings, but their efficiency depends on experimental set-up and, even more importantly, on species or cultivar, and type of explants investigated. A favourable rooting effect of IBA may be due to its higher stability in comparison to IAA, its slow transport from the site of application at the base of cutting and its conversion to IAA in the cutting.

1. Introduction

Acetic Acid

IUPAC name

Acetic acid [1, 2]

Systematic name

Ethanoic acid [3]

Other names

Methanecarboxylic acid [4, 5]

Acetic acid (systematically named ethanoic acid) is an organic compound with the chemical formula CH_3CO_2H (also written as CH_3COOH or $C_2H_4O_2$). It is a colourless liquid which is in undiluted form is also called *glacial acetic acid.* Acetic acid is the main component of vinegar (apart from water; vinegar is roughly 8% acetic acid by volume), and has a distinctive sour taste and pungent smell. Besides its production as household vinegar, it is mainly produced as a precursor to polyvinylacetate and cellulose acetate. Although it is classified as a weak acid, concentrated acetic acid is corrosive, and attacks the skin. Acetic acid is one of the simplest carboxylic acids. It is an important chemical reagent and industrial chemical, mainly used in the production of cellulose acetate mainly for photographic film and polyvinyl acetate for wood glue, as well as for synthetic fibres and fabrics.

In households, diluted acetic acid is often used in descaling agents. In the food industry, acetic acid is used under the food additive code E260 as an acidity regulator and as a condiment. As a food additive it is approved for usage in the EU [6], US [7] and Australia and New Zealand [8].

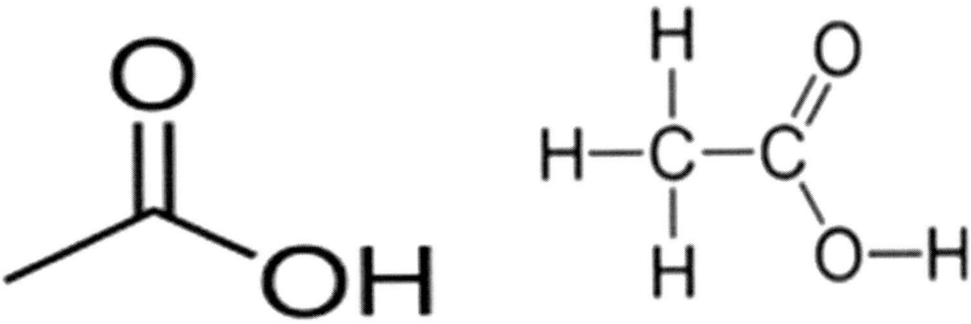

Figure 1. Chemical structure of acetic acid.

The global demand of acetic acid is around 6.5 million tonnes per year (Mt/a), of which approximately 1.5 Mt/a is met by recycling; the remainder is manufactured from petrochemical feedstock [9]. As a chemical reagent, biological sources of acetic acid are of interest but generally uncompetitive. Vinegar is dilute acetic acid, often produced by fermentation and subsequent oxidation of ethanol.

1.1. Nomenclature

The trivial name *acetic acid* is the most commonly used and preferred IUPAC name. The systematic name *ethanoic acid*, a valid IUPAC name, is constructed according to the substitutive nomenclature [10]. The name *acetic acid* derives from *acetum*, the Latin word for vinegar, and is related to the word acid itself.

Glacial acetic acid is a trivial name for water-free (anhydrous) acetic acid. Similar to the German name *Eisessig* (*ice-vinegar*), the name comes from the ice-like crystals that form slightly below room temperature at 16.6 °C (61.9 °F) (the presence of 0.1% water lowers its melting point by 0.2 °C) [11].

A common abbreviation for acetic acid is *AcOH*, where *Ac* stands for the acetyl group CH_3–C(=O)–. Acetate (CH_3COO^-) is abbreviated AcO^-. The *Ac* should not be confused with the abbreviation for the chemical element Actinium. To better reflect its structure, acetic acid is often written as CH_3–C(O)OH, CH_3–C(=O)OH, CH_3COOH, and CH_3CO_2H. In the context of acid-base reactions, the abbreviation *HAc* is sometimes used, where *Ac* stands instead for acetate. Acetate is the ion resulting from loss of H^+ from acetic acid. The name *acetate* can also refer to a salt containing this anion, or an ester of acetic acid.

1.2. History

Vinegar was known early in civilization as the natural result of exposure to air of beer and wine, as acetic acid-producing bacteria are present globally. The use of acetic acid in alchemy extends into the 3rd century BC, when the Greek philosopher Theophrastus described how vinegar acted on metals to produce pigments useful in art, including *white lead* (lead carbonate) and *verdigris*, a green mixture of copper salts including copper (II) acetate. Ancient Romans boiled soured wine to produce a highly sweet syrup called *sapa*. Sapa that was produced in lead pots was rich in lead acetate, a sweet substance also called *sugar of lead* or *sugar of Saturn*, which contributed to lead poisoning among the Roman aristocracy [12]

In the Renaissance, glacial acetic acid was prepared through the dry distillation of certain metal acetates (the most noticeable one being copper (II) acetate). The 16th-century German alchemist Andreas Libavius described such a procedure, and he compared the glacial acetic acid produced by this means to vinegar.

The presence of water in vinegar has such a profound effect on acetic acid's properties that for centuries chemists believed that glacial acetic acid and the acid found in vinegar were two different substances. French chemist Pierre Adet proved them to be identical [12, 13].

In 1847 the German chemist, Hermann Kolbe synthesized acetic acid from inorganic compounds for the first time. This reaction sequence consisted of chlorination of carbon disulfide to carbon tetrachloride, followed by pyrolysis to tetrachloroethylene and aqueous chlorination to trichloroacetic acid, and finally with electrolytic reduction to acetic acid [14].

By 1910, most glacial acetic acid was obtained from the "pyroligneous liquor" from distillation of wood. The acetic acid was isolated from this by treatment with milk of lime, and the resulting calcium acetate was then acidified with sulfuric acid to recover acetic acid. At that time, Germany was producing 10,000 tons of glacial acetic acid, around 30% of which was used for the manufacturing of indigo dye [12, 15].

Because both methanol and carbon monoxide are commodity raw materials, methanol carbonylation long appeared to be an attractive precursor to acetic acid. Henry Dreyfus at British Celanese developed a methanol carbonylation pilot plant as early as 1925 [16]. However, a lack of practical materials that could contain the corrosive reaction mixture at the high pressures needed (200 atm or more) discouraged commercialization of these routes. The first commercial methanol carbonylation process, which used a cobalt catalyst, was developed by the German chemical company BASF in 1963. In 1968, a rhodium-based catalyst (*cis*−$[Rh(CO)_2I_2]^-$) was discovered that could operate efficiently at lower pressure with almost no by-products. The US chemical company built the first plant using this catalyst in 1970, and rhodium-catalysed methanol carbonylation became the dominant method of acetic acid production (see Monsanto process). In the late 1990s, the chemical company BP commercialized the Cativa catalyst ($[Ir(CO)_2I_2]^-$), which is promoted by ruthenium for greater efficiency. This iridium-catalysed Cativa process is greener and more efficient [17] and has largely replaced the Monsanto process, often in the same production plants.

1.2.1. In the Interstellar Medium

Acetic acid was discovered in the interstellar medium in 1996 by a team led by David Mehringer [18] who detected it using the former Berkeley-Illinois-Maryland Association array at the Hat Creek Radio Observatory and the former Millimeter Array located at the Owens Valley Radio Observatory. It was first detected in the Sagittarius B2 North molecular cloud (also known as the Sgr B2 Large Molecule Heimat source). Acetic acid has the distinction of being the first molecule discovered in the interstellar medium using solely radio interferometers; in all previous ISM molecular discoveries made in the millimeter and centimeter wavelength regimes, single dish radio telescopes were at least partly responsible for the detections.

1.3. Chemical Properties

Molecular formula: $C_2H_4O_2$
Molar mass: 60.05 g mol^{-1}
Appearance: Colourless liquid
Density: 1.049 g cm^{-3}
Melting point: 16-17 °C, 289-290 K, 61-62 °F
Boiling point: 118-119 °C, 391-392 K, 244-246 °F
Solubility in water: Miscible
log P: -0.322
Acidity (pK_a): 4.76
Basicity (pK_b): 9.198
Viscosity: 1.22 mPa s
Dipole moment: 1.74 D

1.3.1. Acidity

The hydrogen center in the carboxyl group (−COOH) in carboxylic acids, such as acetic acid, can separate from the molecule by ionization:

$$CH_3CO_2H \rightarrow CH_3CO_2^- + H^+$$

Because of this release of the proton (H^+), acetic acid has acidic character. Acetic acid is a weak monoprotic acid. In aqueous solution, it has a pK_a value of 4.75. Its conjugate base is acetate (CH_3COO^-). A 1.0 M solution (about the concentration of domestic vinegar) has a pH of 2.4, indicating that merely 0.4% of the acetic acid molecules are dissociated.

1.3.2. Structure

In solid acetic acid, the molecules form pairs (dimers), being connected by hydrogen bonds [19]. The dimers can also be detected in the vapour at 120 °C (248 °F). Dimers also occur in the liquid phase in dilute solutions in non-hydrogen-bonding solvents, and a certain extent in pure acetic acid [20], but are disrupted by hydrogen-bonding solvents. The dissociation enthalpy of the dimer is estimated at 65.0–66.0 kJ/mol, and the dissociation entropy at 154–157 J mol^{-1} K^{-1} [21]. Other lower carboxylic acids dimerize in a similar fashion.

1.3.3. Solvent Properties

Liquid acetic acid is a hydrophilic (polar) protic solvent, similar to ethanol and water. With a moderate relative static permittivity (dielectric constant) of 6.2, it dissolves not only polar compounds such as inorganic salts and sugars, but also non-polar compounds such as oils and elements such as sulfur and iodine.

It readily mixes with other polar and non-polar solvents such as water, chloroform, and hexane. With higher alkanes (starting with octane), acetic acid is not completely miscible anymore, and its miscibility continues to decline with longer n-alkanes [22]. This dissolving property and miscibility of acetic acid makes it a widely used industrial chemical. Its solvent properties are mainly of value in the production of dimethyl terephthalate [9].

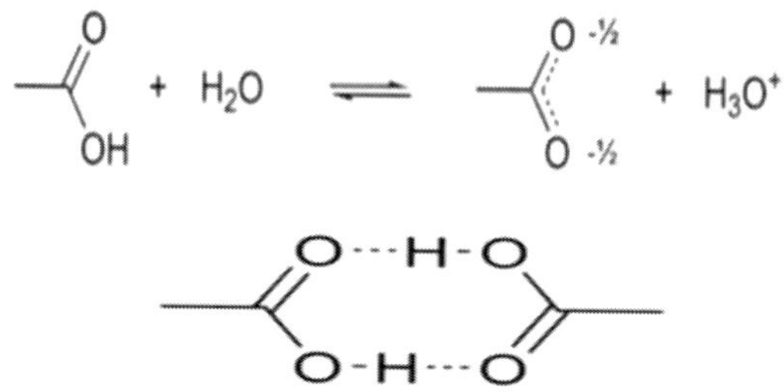

Figure 2. Cyclic dimer of acetic acid; dashed lines represent hydrogen bonds.

1.4. Chemical Reactions

1.4.1. Organic Chemistry

Acetic acid undergoes the typical chemical reactions of a carboxylic acid. Upon treatment with a standard base, it converts to metal acetate and water. With strong bases (e.g., organolithium reagents), it can be doubly deprotonated to give $LiCH_2CO_2Li$. Reduction of acetic acid gives ethanol. The OH group is the main site of reaction, as illustrated by the conversion of acetic acid to acetyl chloride.

Other substitution derivatives include acetic anhydride; this anhydride is produced by loss of one molecule of water from two molecules of acetic acid. Esters of acetic acid can likewise be formed via Fischer esterification, and also amides. When heated above 440 °C (824 °F), acetic acid decomposes to produce carbon dioxide and methane, or to produce ketene and water.

1.4.2. Reactions with Inorganic Compounds

Acetic acid is mildly corrosive to metals including iron, magnesium, and zinc, forming hydrogen gas and salts called acetates:

$$Mg + 2\ CH_3COOH \rightarrow (CH_3COO)_2Mg + H_2$$

Because aluminium forms a passivating acid-resistant film of aluminium oxide, aluminium tanks are used to transport acetic acid.

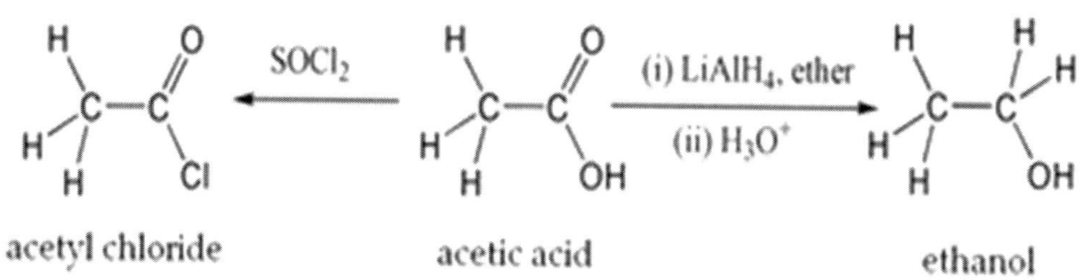

Figure 3. Chemical reactions of acetic acid.

Metal acetates can also be prepared from acetic acid and an appropriate base, as in the popular "baking soda + vinegar" reaction:

$$NaHCO_3 + CH_3COOH \rightarrow CH_3COONa + CO_2 + H_2O$$

A colour reaction for salts of acetic acid is iron (III) chloride solution, which results in a deeply red colour that disappears after acidification. Acetates when heated with arsenic trioxide form cacodyl oxide, which can be detected by its malodorous vapours.

1.4.3. Biochemistry

At physiological pHs, acetic acid is usually fully ionized to acetate. In biochemistry, acetate and acetic acid are equivalent. The acetyl group, derived from acetic acid, is fundamental to all forms of life. When bound to coenzyme A, it is central to the metabolism of carbohydrates and fats. Unlike longer-chain carboxylic acids (the fatty acids), acetic acid does not occur in natural triglycerides. However, the artificial triglyceride triacetin (glycerine triacetate) is a common food additive and is found in cosmetics and medicines.

Acetic acid is produced and excreted by acetic acid bacteria, notable ones being the *Acetobacter* genus and *Clostridium acetobutylicum*. These bacteria are found universally in foodstuffs, water, and soil, and acetic acid is produced naturally as fruits and other foods spoil. Acetic acid is also a component of the vaginal lubrication of humans and other primates, where it appears to serve as a mild antibacterial agent [23].

1.5. Production

Acetic acid is produced industrially both synthetically and by bacterial fermentation. About 75% of acetic acid produced in the chemical industry is made by the carbonylation of methanol, explained below [9]. Alternative methods account for the rest. The biological route accounts for only about 10% of world production, but it remains important for the production of vinegar, as many food purity laws stipulate that vinegar used in foods must be of biological origin. As of 2003 - 2005, total worldwide production of virgin acetic acid was estimated at 5 Mt/a (million tonnes per year), approximately half of which was then produced in the United States. European production stood at approximately 1 Mt/a and was declining, and 0.7 Mt/a were produced in Japan. Another 1.5 Mt were recycled each year, bringing the total world market to 6.5 Mt/a [24, 25]. Since then the global production has increased to 10.7 Mt/a (in 2010), and further, however, a slowing increase in production is predicted [26]. The two biggest producers of virgin acetic acid are Celanese and BP Chemicals. Other major producers include Millennium Chemicals, Sterling Chemicals, Samsung, Eastman, and Svensk Etanolkemi.

1.5.1. Methanol Carbonylation

Most acetic acid is produced by methanol carbonylation. In this process, methanol and carbon monoxide react to produce acetic acid according to the equation:

$$CH_3OH + CO \rightarrow CH_3COOH$$

The process involves iodomethane as an intermediate, and occurs in three steps. A catalyst, metal carbonyl, is needed for the carbonylation (step 2) [27].

1 $CH_3OH + HI \rightarrow CH_3I + H_2O$
2 $CH_3I + CO \rightarrow CH_3COI$
3 $CH_3COI + H_2O \rightarrow CH_3COOH + HI$

By altering the process conditions, acetic anhydride may also be produced on the same plant.

1.5.2. Acetaldehyde Oxidation

Prior to the commercialization of the Monsanto process, most acetic acid was produced by oxidation of acetaldehyde. This remains the second-most-important manufacturing method, although it is usually uncompetitive with the carbonylation of methanol.

The acetaldehyde may be produced via oxidation of butane or light naphtha, or by hydration of ethylene. When butane or light naphtha is heated with air in the presence of various metal ions, including those of manganese, cobalt, and chromium, peroxides form and then decompose to produce acetic acid according to the chemical equation:

$$2\ C_4H_{10} + 5\ O_2 \rightarrow 4\ CH_3COOH + 2\ H_2O$$

The typical reaction is conducted at temperatures and pressures designed to be as hot as possible while still keeping the butane a liquid. Typical reaction conditions are 150 °C (302 °F) and 55 atm. Side-products may also form, including butanone, ethyl acetate, formic acid, and propionic acid. These side-products are also commercially valuable, and the reaction conditions may be altered to produce more of them where needed. However, the separation of acetic acid from these by-products adds to the cost of the process.

Under similar conditions and using similar catalysts, as are used for butane oxidation, the oxygen in air to produce acetic acid can oxidize acetaldehyde.

$$2\ CH_3CHO + O_2 \rightarrow 2\ CH_3COOH$$

Using modern catalysts, this reaction can have an acetic acid yield greater than 95%. The major side-products are ethyl acetate, formic acid, and formaldehyde, all of which have lower boiling points than acetic acid and are readily separated by distillation [28].

1.5.3. Ethylene Oxidation

Acetaldehyde may be prepared from ethylene via the Wacker process, and then oxidized as above. In more recent times, chemical company Showa Denko, which opened an ethylene oxidation plant in Ōita, Japan, in 1997, commercialized a cheaper single-stage conversion of ethylene to acetic acid [29].

The process is catalysed by a palladium metal catalyst supported on a heteropoly acid such as tungstosilicic acid. It is thought to be competitive with methanol carbonylation for smaller plants (100–250 kt/a), depending on the local price of ethylene.

1.5.4. Oxidative Fermentation

For most of human history, acetic acid bacteria of the genus *Acetobacter* have made acetic acid, in the form of vinegar. Given sufficient oxygen, these bacteria can produce vinegar from a variety of alcoholic foodstuffs. Commonly used feeds include apple cider, wine, and fermented grain, malt, rice, or potato mashes. The overall chemical reaction facilitated by these bacteria is:

$$C_2H_5OH + O_2 \rightarrow CH_3COOH + H_2O$$

A dilute alcohol solution inoculated with *Acetobacter* and kept in a warm, airy place will become vinegar over the course of a few months. Industrial vinegar-making methods accelerate this process by improving the supply of oxygen to the bacteria.

The first batches of vinegar produced by fermentation probably followed errors in the winemaking process. If must is fermented at too high a temperature, acetobacter will overwhelm the yeast naturally occurring on the grapes. As the demand for vinegar for culinary, medical, and sanitary purposes increased, vintners quickly learned to use other organic materials to produce vinegar in the hot summer months before the grapes were ripe and ready for processing into wine. This method was slow, however, and not always successful, as the vintners did not understand the process [30]. One of the first modern commercial processes was the "fast method" or "German method", first practised in Germany in 1823. In this process, fermentation takes place in a tower packed with wood shavings or charcoal. The alcohol-containing feed is trickled into the top of the tower, and fresh air supplied from the bottom by either natural or forced convection. The improved air supply in this process cut the time to prepare vinegar from months to weeks [31]. Nowadays, most vinegar is made in submerged tank culture, first described in 1949 by Otto Hromatka and Heinrich Ebner [32]. In this method, alcohol is fermented to vinegar in a continuously stirred tank, and oxygen is supplied by bubbling air through the solution. Using modern applications of this method, vinegar of 15% acetic acid can be prepared in only 24 hours in batch process, even 20% in 60-hour fed-batch process [30].

1.5.5. Anaerobic Fermentation

Species of anaerobic bacteria, including members of the genus *Clostridium* or *Acetobacterium* can convert sugars to acetic acid directly, without using ethanol as an intermediate. The overall chemical reaction conducted by these bacteria may be represented as:

$$C_6H_{12}O_6 \rightarrow 3\ CH_3COOH$$

These acetogenic bacteria produce acetic acid from one-carbon compounds, including methanol, carbon monoxide, or a mixture of carbon dioxide and hydrogen:

$$2\ CO_2 + 4\ H_2 \rightarrow CH_3COOH + 2\ H_2O$$

This ability of *Clostridium* to utilize sugars directly, or to produce acetic acid from less costly inputs, means that these bacteria could potentially produce acetic acid more efficiently than ethanol-oxidizers like *Acetobacter*.

However, *Clostridium* bacteria are less acid-tolerant than *Acetobacter*. Even the most acid-tolerant *Clostridium* strains can produce vinegar of only a few per cent acetic acid, compared to *Acetobacter* strains that can produce vinegar of up to 20% acetic acid. At present, it remains more cost-effective to produce vinegar using *Acetobacter* than to produce it using *Clostridium* and then concentrate it. As a result, although acetogenic bacteria have been known since 1940, their industrial use remains confined to a few niche applications [33].

1.6. Uses

Acetic acid is a chemical reagent for the production of chemical compounds. The largest single use of acetic acid is in the production of vinyl acetate monomer, closely followed by acetic anhydride and ester production. The volume of acetic acid used in vinegar is comparatively small [25].

1.6.1. Vinyl Acetate Monomer

The major use of acetic acid is for the production of vinyl acetate monomer (VAM). This application consumes approximately 40% to 45% of the world's production of acetic acid. The reaction is of ethylene and acetic acid with oxygen over a palladium catalyst.

$$2\ H_3C\text{–}COOH + 2\ C_2H_4 + O_2 \rightarrow 2\ H_3C\text{–}CO\text{–}O\text{–}CH{=}CH_2 + 2\ H_2O$$

Vinyl acetate can be polymerized to polyvinyl acetate or to other polymers, which are components in paints and adhesives.

1.6.2. Ester Production

The major esters of acetic acid are commonly used solvents for inks, paints and coatings. The esters include ethyl acetate, n-butyl acetate, isobutyl acetate, and propyl acetate. They are typically produced by catalysed reaction from acetic acid and the corresponding alcohol:

$$H_3C\text{-}COOH + HO\text{-}R \rightarrow H_3C\text{-}CO\text{-}O\text{-}R + H_2O,\ (R = \text{a general alkyl group})$$

Most acetate esters, however, are produced from acetaldehyde using the Tishchenko reaction. In addition, ether acetates are used as solvents for nitrocellulose, acrylic lacquers, varnish removers, and wood stains. First, glycol monoethers are produced from ethylene oxide or propylene oxide with alcohol, which are then esterified with acetic acid. The three major products are ethylene glycol monoethyl ether acetate (EEA), ethylene glycol monobutyl ether acetate (EBA), and propylene glycol monomethyl ether acetate (PMA, more commonly known as PGMEA in semiconductor manufacturing processes). This application consumes about 15% to 20% of worldwide acetic acid. Ether acetates, for example EEA, have been shown to be harmful to human reproduction [25].

1.6.3. Acetic Anhydride

The product of the condensation of two molecules of acetic acid is acetic anhydride. The worldwide production of acetic anhydride is a major application, and uses approximately 25%

to 30% of the global production of acetic acid. The main process involves dehydration of acetic acid to give ketene, which condenses with acetic acid to give the anhydride:

$$CH_3CO_2H \rightarrow CH_2{=}C{=}O + H_2O$$

$$CH_3CO_2H + CH_2{=}C{=}O \rightarrow (CH_3CO)_2O$$

Acetic anhydride is an acetylation agent. As such, its major application is for cellulose acetate, a synthetic textile also used for photographic film. Acetic anhydride is also a reagent for the production of heroin and other compounds.

1.6.4. Vinegar

Vinegar is typically 4-18% acetic acid by mass. Vinegar is used directly as a condiment, and in the pickling of vegetables and other foods. Table vinegar tends to be more diluted (4% to 8% acetic acid), while commercial food pickling employs solutions that are more concentrated. The amount of acetic acid used as vinegar on a worldwide scale is not large, but is by far the oldest and best-known application.

1.6.5. Use As Solvent

Glacial acetic acid is an excellent polar protic solvent, as noted above. It is frequently used as a solvent for recrystallization to purify organic compounds. Acetic acid is used as a solvent in the production of terephthalic acid (TPA), the raw material for polyethylene terephthalate (PET). In 2006, about 20% of acetic acid is used for TPA production [25].

Acetic acid is often used as a solvent for reactions involving carbocations, such as Friedel-Crafts alkylation. For example, one stage in the commercial manufacture of synthetic camphor involves a Wagner-Meerwein rearrangement of camphene to isobornyl acetate; here acetic acid acts both as a solvent and as a nucleophile to trap the rearranged carbocation. Acetic acid is the solvent of choice when reducing an aryl nitro-group to aniline using palladium-on-carbon.

Glacial acetic acid is used in analytical chemistry for the estimation of weakly alkaline substances such as organic amides. Glacial acetic acid is a much weaker base than water, so the amide behaves as a strong base in this medium. It then can be titrated using a solution in glacial acetic acid of a very strong acid, such as perchloric acid.

1.6.6. Niche Applications

Dilute solutions of acetic acids are also used as a stop bath during the development of photographic films, and in descaling agents to remove limescale from taps and kettles. In the clinical laboratory dilute acetic acid lyse red blood cells in order to facilitate microscopic examination. Acetic acid in the form of household vinegar is often used to clean indoor climbing holds of chalk (magnesium carbonate). The acidity is also used for treating the sting of the box jellyfish by disabling the stinging cells of the jellyfish, preventing serious injury or death if applied immediately, and for treating outer ear infections in people in preparations such as Vosol. In this manner, acetic acid is used as a spray-on preservative for livestock silage, to discourage bacterial and fungal growth. Acetic acid vapour is used for the fumigation of bee hives to control *Nosema* and other pests [34].

Glacial acetic acid is also used as a wart and verruca remover. Organic or inorganic salts are produced from acetic acid, including:

- Sodium acetate, used in the textile industry and as a food preservative (E262).
- Copper(II) acetate, used as a pigment and a fungicide.
- Aluminium acetate and iron(II) acetate—used as mordants for dyes.
- Palladium(II) acetate, used as a catalyst for organic coupling reactions such as the Heck reaction.
- Silver acetate, used as a pesticide.
- Substituted acetic acids produced include:
- Monochloroacetic acid (MCA), dichloroacetic acid (considered a by-product), and trichloroacetic acid. MCA is used in the manufacture of indigo dye.
- Bromoacetic acid, which is esterified to produce the reagent ethyl bromoacetate.
- Trifluoroacetic acid, which is a common reagent in organic synthesis.

Amounts of acetic acid used in these other applications together (apart from TPA) account for another 5–10% of acetic acid use worldwide. These applications are, however, not expected to grow as much as TPA production [25]. Diluted acetic acid is also used in physical therapy to break up nodules of scar tissue via iontophoresis.

2. Auxins

Auxins are a class of plant hormones (or plant growth substances) with some morphogen-like characteristics. Auxins have a cardinal role in coordination of many growth and behavioral processes in the plant's life cycle and are essential for plant body development. Auxins and their role in plant growth were first described by the Dutch scientist Frits Went [35]. Kenneth V. Thimann isolated this phytohormone and determined its chemical structure as indole-3-acetic acid. Went and Thimann then co-authored a book on plant hormones, *Phytohormones*, in 1937.

Native auxins. indole-3-acetic acid (IAA) is the most abundant and the basic auxin natively occurring and functioning in plants. It generates the majority of auxin effects in intact plants, and is the most potent native auxin.

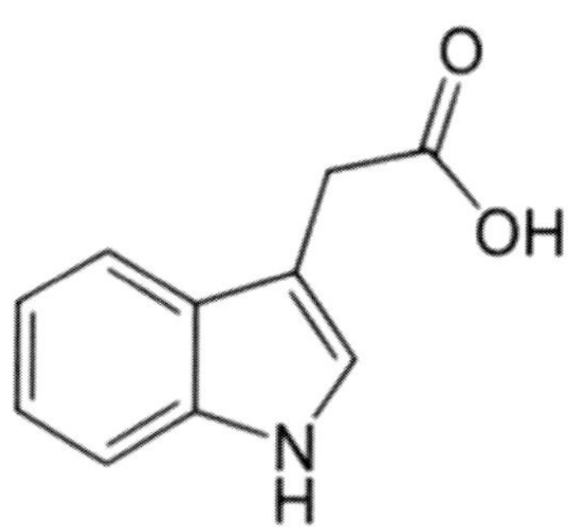

Figure 4. Chemical structure of indole-3-acetic acid (IAA).

There are three more native - endogenous auxins [36]. All auxins are compounds with aromatic ring and a carboxylic acid group: [37]

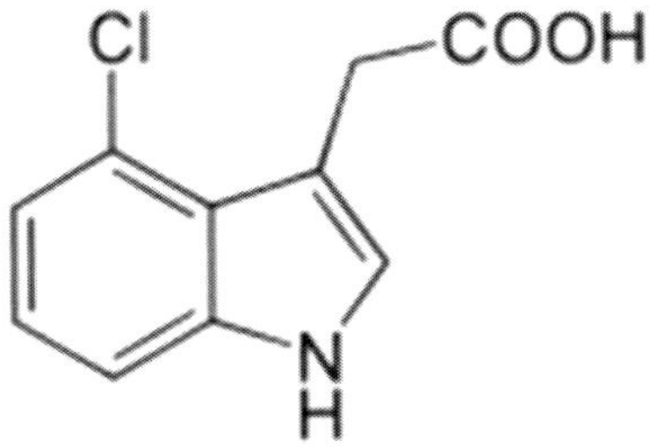

Figure 5. Chemical structure of 4-chloroindole-3-acetic acid (4-CI-IAA).

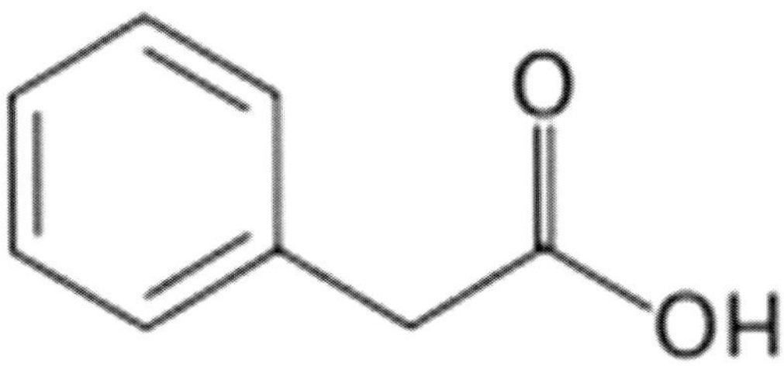

Figure 6. Chemical structure of 2-phenylacetic acid (PAA).

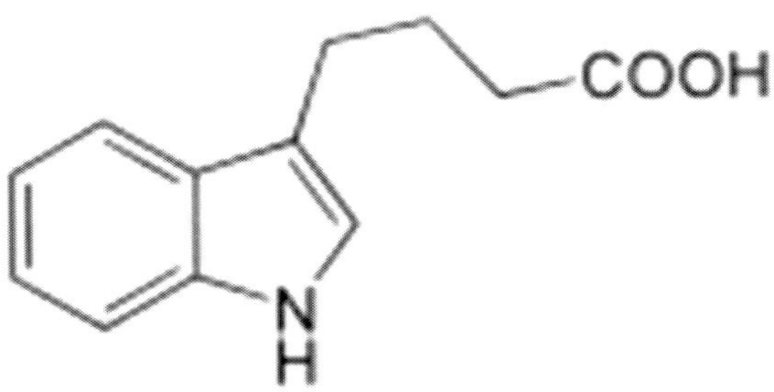

Figure 7. Chemical structure of indole-3-butyric acid (IBA).

2.1. Overview

Auxins derive their name from the Greek word *auxein* - "to grow/increase". They were the first of the major plant hormones to be discovered.

The (dynamic and environment responsive) pattern of auxin distribution within the plant is a key factor for plant growth, its reaction to its environment, and specifically for development of plant organs [38, 39] (such as leaves or flowers). It is achieved through very complex and well coordinated active transport of auxin molecules from cell to cell throughout the plant body-by the so-called polar auxin transport [38].

Thus, a plant can (as a whole) react to external conditions and adjust to them, without requiring a nervous system. Auxins typically act in concert with, or in opposition to, other plant hormones. For example, the ratio of auxin to cytokinin in certain plant tissues determines initiation of root versus shoot buds.

On the molecular level, all auxins are compounds with an aromatic ring and a carboxylic acid group [3]. The most important member of the auxin family is indole-3-acetic acid (IAA) [36]. IAA generates the majority of auxin effects in intact plants, and is the most potent native auxin. And as a native auxin, its stability is controlled in many ways in plants, from synthesis, through possible conjugation to degradation of its molecules, always according to the requirements of the situation. However, molecules of IAA are chemically labile in aqueous solution, so it is not used commercially as a plant growth regulator.

The four naturally occurring (endogenous) auxins are IAA, 4-chloroindole-3-acetic acid, phenylacetic acid and indole-3-butyric acid; only these four were found to be synthesized by plants [36].

However, most of the knowledge described so far in auxin biology and as described in the article below, apply basically to IAA; the other three endogenous auxins seems to have rather marginal importance for intact plants in natural environments. Alongside endogenous auxins, scientists and manufacturers have developed many synthetic compounds with auxinic activity.

Synthetic auxin analogs include 1-naphthaleneacetic acid, 2,4-dichlorophenoxyacetic acid (2,4-D), [36] and many others. Some synthetic auxins, such as 2,4-D and 2,4,5-trichlorophenoxyacetic acid (2,4,5-T), are used also as herbicides. Broad-leaf plants (dicots), such as dandelions, are much more susceptible to auxins than narrow-leaf plants (monocots) such as grasses and cereal crops, so these synthetic auxins are valuable as synthetic herbicides.

Auxins are also often used to promote initiation of adventitious roots, and are the active ingredient of the commercial preparations used in horticulture to root stem cuttings. They can also be used to promote uniform flowering and fruit set, and to prevent premature fruit drop.

2.2. Hormonal Activity

Auxins coordinate development at all levels in plants, from the cellular level, through organs, and ultimately to the whole plant.The plant cell wall is made up of cellulose, proteins, and, in many cases, lignin. It is very firm and prevents any sudden expansion of cell volume (and, without the contribution of auxins, any expansion at all).

2.3. Molecular Mechanisms

Auxin molecules present in cells may trigger responses directly through stimulation or inhibition of the expression of sets of certain genes [40] or by means independent of gene expression. Auxin transcriptionally activates four different families of *early genes* (aka *primary response genes*), so-called because the components required for the activation are preexisting, leading to a rapid response [41]. The families are glutathione S-transferases, auxin homeostasis proteins like GH3, SAUR genes of currently unknown function, and the Aux/IAA repressors [42].

One of the pathways leading to the changes of gene expression involves the reception of auxin by TIR1 protein. In 2005, the F-box protein TIR1, which is part of the ubiquitin ligase complex SCF^{TIR1}, was demonstrated to be an auxin receptor [43]. Upon binding of auxin, TIR1 recruits specific transcriptional repressors (the Aux/IAA repressors) for ubiquitination by the SCF complex.

This marking process leads to the degradation of the Aux/IAAs repressors by the proteasome. The degradation of the repressors leads, in turn, to potentiation of auxin response factor-mediated transcription of specific genes in response to auxins [44]. Another protein, auxin-binding protein 1 (ABP1), is a putative receptor for different signaling pathway, but its role is as yet unclear. Electrophysiological experiments with protoplasts and anti-ABP1 antibodies suggest ABP1 may have a function at the plasma membrane, and cells can possibly use ABP1 proteins to respond to auxin through means faster and independent of gene expression.

2.3.1. On a Cellular Level

On the cellular level, auxin is essential for cell growth, affecting both cell division and cellular expansion. Auxin concentration level, together with other local factors, contributes to cell differentiation and specification of the cell fate.

Depending on the specific tissue, auxin may promote axial elongation (as in shoots), lateral expansion (as in root swelling), or isodiametric expansion (as in fruit growth). In some cases (coleoptile growth), auxin-promoted cellular expansion occurs in the absence of cell division.

In other cases, auxin-promoted cell division and cell expansion may be closely sequenced within the same tissue (root initiation, fruit growth). In a living plant, auxins and other plant hormones nearly always appear to interact to determine patterns of plant development.

2.3.2. Organ Patterns

Growth and division of plant cells together result in growth of tissue, and specific tissue growth contributes to the development of plant organs.

Growth of cells contributes to the plant's size, unevenly localized growth produces bending, turning and directionalization of organs- for example, stems turning toward light sources (phototropism), roots growing in response to gravity (gravitropism), and other tropisms originated because cells on one side grow faster than the cells on the other side of the organ. So, precise control of auxin distribution between different cells has paramount importance to the resulting form of plant growth and organization.

2.3.4. Uneven Distribution of Auxin

To cause growth in the required domains, auxins must of necessity be active preferentially in them. Auxins are not synthesized in all cells (even if cells retain the potential ability to do so, only under specific conditions will auxin synthesis be activated in them). For that purpose, auxins have to be not only translocated toward those sites where they are needed, but also they must have an established mechanism to detect those sites. For that purpose, auxins have to be translocated toward those sites where they are needed. Translocation is driven throughout the plant body, primarily from shoot apex to root apex (from up to down).

For long distances, relocation occurs via the stream of fluid in phloem vessels, but, for short-distance transport, a unique system of coordinated polar transport directly from cell to cell is exploited. This short-distance, active transport exhibits some morphogenetic properties.

This process, the polar auxin transport, is directional, very strictly regulated, and based in uneven distribution of auxin efflux carriers on the plasma membrane, which send auxins in the proper direction. Pin-formed (PIN) proteins are vital in transporting auxin [39, 45].

The regulation of PIN protein localisation in a cell determines the direction of auxin transport from cell, and concentrated effort of many cells creates peaks of auxin, or auxin maxima (regions having cells with higher auxin - a maximum) [39]. Proper and timely auxin maxima within developing roots and shoots are necessary to organise the development of the organ [38, 45-46]. Surrounding auxin maxima are cells with low auxin troughs, or auxin minima. For example, in the *Arabidopsis* fruit, auxin minima have been shown to be important for its tissue development [48].

2.3.5. Organization of the Plant

As auxins contribute to organ shaping [38-39], they are also fundamentally required for proper development of the plant itself. [4] Without hormonal regulation and organization, plants would be merely proliferating heaps of similar cells. Auxin employment begins in the embryo of the plant, where directional distribution of auxin ushers in subsequent growth and development of primary growth poles, then forms buds of future organs. Next, it helps to coordinate proper development of the arising organs, such as roots, cotyledons and leaves and mediates long distance signals between them, contributing so to the overal architecture of the plant [38]. Throughout the plant's life, auxin helps the plant maintain the polarity of growth, [38] and actually "recognize" where it has its branches (or any organ) connected.

An important principle of plant organization based upon auxin distribution is apical dominance, which means the auxin produced by the apical bud (or growing tip) diffuses (and is transported) downwards and inhibits the development of ulterior lateral bud growth, which would otherwise compete with the apical tip for light and nutrients. Removing the apical tip and its suppressively acting auxin allows the lower dormant lateral buds to develop, and the buds between the leaf stalk and stem produce new shoots which compete to become the leading growth. The process is actually quite complex, because auxin transported downwards from the lead shoot tip has to interact with several other plant hormones (such as strigolactones or cytokinins) in the process on various positions along the growth axis in plant body to achieve this phenomenon. This plant behavior is used in pruning by horticulturists.

Finally, the sum of auxin arriving from stems to roots influences the degree of root growth. If shoot tips are removed, the plant does not react just by outgrowth of lateral buds-which are supposed to replace to original leading.

It also follows that smaller amount of auxin arriving to the roots results in slower growth of roots and the nutrients are subsequently in higher degree invested in the upper part of the plant, which hence starts to grow faster.

2.4. Effects

Crown galls are caused by *Agrobacterium tumefaciens* bacteria; they produce and excrete auxin and cytokinin, which interfere with normal cell division and cause tumors.

Auxin participates in phototropism, geotropism, hydrotropism and other developmental changes. The uneven distribution of auxin, due to environmental cues, such as unidirectional light or gravity force, results in uneven plant tissue growth, and generally, auxin governs the form and shape of plant body, direction and strength of growth of all organs, and their mutual interaction [39]. Auxin stimulates cell elongation by stimulating wall-loosening factors, such as elastins, to loosen cell walls. The effect is stronger if gibberellins are also present. Auxin also stimulates cell division if cytokinins are present. When auxin and cytokinin are applied to callus, rooting can be generated if the auxin concentration is higher than cytokinin concentration. Xylem tissues can be generated when the auxin concentration is equal to the cytokinins.

Auxin also induces sugar and mineral accumulation at the site of application.

2.4.1. Wound Response

Auxin induces the formation and organization of phloem and xylem. When the plant is wounded, auxin may induce cell differentiation and regeneration of the vascular tissues.

2.4.2. Root Growth and Development

Auxins promote root initiation [49]. Auxin induces both growth of pre-existing roots and adventitious root formation, i.e., branching of the roots. As more native auxin is transported down the stem to the roots, the overall development of the roots is stimulated. If the source of auxin is removed, for example the tips of stems are trimmed, the roots are less stimulated accordingly, and growth of stem is supported instead. In horticulture, auxins, especially NAA and IBA, are commonly applied to stimulate root initiation of cuttings. However, high concentrations of auxin inhibit root elongation. Removal of the root tip can lead to inhibition of secondary root formation.

2.4.3. Apical Dominance

Auxin induces shoot apical dominance; the axillary buds are inhibited by auxin, as a high concentration of auxin directly stimulates ethylene synthesis in lateral buds, causing inhibition of their growth and potentiation of apical dominance. When the apex of the plant is removed, the inhibitory effect is removed and the growth of lateral buds is enhanced. Auxin is sent to the part of the plant facing light, and this promotes growth towards that direction [50].

2.4.4. Fruit Growth and Development

Auxin is required for fruit growth and development and delays fruit senescence. When seeds are removed from strawberries, fruit growth is stopped; exogenous auxin stimulates the growth in fruits with seeds removed. For fruit with unfertilized seeds, exogenous auxin results in parthenocarpy ("virgin-fruit" growth). Fruits form abnormal morphologies when auxin transport is disturbed [51]. In *Arabidopsis* fruits, auxin controls the release of seeds from the fruit (pod). The valve margins are a specialised tissue in pods that regulates pod dehiscence. Auxin must be removed from the valve margin cells to allow the valve margins to form. This process requires modification of the auxin transporters (PIN proteins) [48].

2.4.5. Flowering

Auxin plays also a minor role in the initiation of flowering and development of reproductive organs. In low concentrations, it can delay the senescence of flowers.

2.4.6. Ethylene Biosynthesis

In low concentrations, auxin can inhibit ethylene formation and transport of precursor in plants; however, high concentrations can induce the synthesis of ethylene. Therefore, the high concentration can induce femaleness of flowers in some species.

Auxin inhibits abscission prior to formation of abscission layer, and thus inhibits senescence of leaves.

3. Indole-3-Acetic Acid (IAA)

Indole-3-acetic acid (IAA) is the most common, naturally-occurring, plant hormone of the auxin class. It is the best known of the auxins, and has been the subject of extensive studies by plant physiologists [52]. Chemically, IAA is a carboxylic acid in which the carboxyl group is attached through a methylene group to the C-3 position of an indole ring. In appearance, IAA is a colorless solid.

IUPAC name
2-(1*H*-indol-3-yl) acetic acid
Other names
Indole-3-acetic acid
Indolylacetic acid
1H-Indole-3-acetic acid
Indoleacetic acid
heteroauxin
IAA

Chemical properties IAA
Molecular formula: $C_{10}H_9NO_2$
Molar mass: 175.184
Appearance: white solid
Melting point: 168-170 °C (441-443 K)
Solubility in water: moderate

3.1. Biosynthesis and Biological Activity

IAA is predominantly produced in cells of the apex (bud) and very young leaves of a plant. Plants can synthesize IAA by several independent biosynthetic pathways. Four of them start from tryptophan, but there is also a biosynthetic pathway independent of tryptophan [53] Plants mainly produce IAA from tryptophan through indole-3-pyruvic acid [54, 55] IAA is also produced from tryptophan through indole-3-acetaldoxime in Arabidopsis [56].

3.2. Synthesis

Chemically, it can be synthesized by the reaction of indole with glycolic acid in the presence of a base at 250 °C:[57].

Many methods for its synthesis have been developed since its original synthesis from indole-3-acetonitrile [58].

4. Indole-3-Butyric Acid (IBA)

Indole-3-butyric acid (1H-Indole-3-butanoic acid, IBA) is a white to light-yellow crystalline solid, with the molecular formula $C_{12}H_{13}NO_2$. It melts at 125 °C in atmospheric pressure and decomposes before boiling.

IUPAC name
1H-Indole-3-butanoic-acid
Other names
indole-3-butyric-acid
3-indolebutyric-acid
indolebutyric-acid
IBA

4.1. Plant Hormone

IBA is a plant hormone in the auxin family and is an ingredient in many commercial horticultural plant rooting products.

Since IBA is not soluble in water, it is typically dissolved in 50% or pure alcohol for use in plant rooting, making a solution of between 10,000 to 50,000 ppm. This alcohol solution is then diluted with distilled water to the desired concentration. IBA is also available as a salt, which is soluble in water.

The solution should be kept in a cool, dark place for best results.

This compound had been thought to be strictly synthetic; however, it was reported that the compound was isolated from leaves and seeds of maize and other species. This chemical may also be extracted from the Salix (Willow) genus [59].

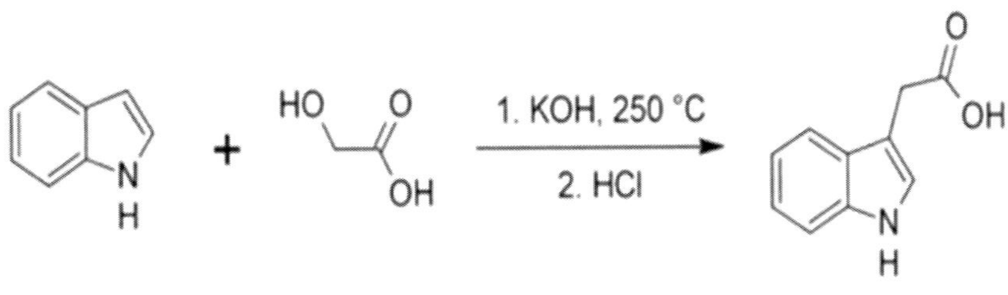

Figure 8. Synthesis reaction of IAA.

5. 1-Naphthaleneacetic Acid (NAA)

1-Naphthaleneacetic acid (NAA) is an organic compound with the formula $C_{10}H_7CH_2$ CO_2H. This colorless solid is soluble in organic solvents. It features a carboxylmethyl group (CH_2CO_2H) linked to the "1-position" of naphthalene.

IUPAC name
2-(1-Naphthyl)acetic acid
Other names
1-Naphthaleneacetic acid
α-Naphthaleneacetic acid
Naphthylacetic acid

Chemical properties [60]
Molecular formula: $C_{12}H_{10}O_2$
Molar mass: 186.21 g mol^{-1}
Appearance: White powder
Melting point: 135 °C (275 °F)
Solubility in water: 0.38 g/L (17 °C)
Acidity (pK_a): 4.24

5.1. Use and Regulation

NAA is a synthetic plant hormone in the auxin family and is an ingredient in many commercial plant rooting horticultural products; it is a rooting agent and used for the vegetative propagation of plants from stem and leaf cutting. It is also used for plant tissue culture [61].

5.2. Specific Uses of NAA

NAA is widely used in agriculture for various purposes. It is considered to be only slightly toxic but when at higher concentrations it can be toxic to animals. This was shown when tested on rats via oral ingestion at 1000-5900 mg/kg [62]. NAA has been shown to greatly increase cellulose fiber formation in plants when paired with another phytohormone called gibberellic acid [63]. Because it is in the auxin family it has also been understood to prevent premature dropping and thinning of fruits from stems. It is applied after blossom fertilization. Increased amounts of it can actually have negative effects however, and cause growth inhibition to the development of plant crops. It has been used on many different crops like apples, olives, oranges, potatoes, and various other fruits. In order for it to obtain its desired effects it must be applied in concentrations ranging from 20 to 100 μg/mL [64]. In micro propagation of various plants NAA is typically added to a media containing nutrients essential to the plants survival. It is added to help induce root formation in various plant types.

It can also be applied by spraying it on to plants and which is typical for agricultural use. It is prohibited in many areas to use it in high concentrations due to the health concerns towards humans and other animals [65, 66].

5.3. How to Detect NAA

Because of the potential hazards to humans NAA presents, methods are needed to detect it. Some of these methods include Liquid Chromatography-Ultra Violet Detection (LC-UVD), High Performance Liquid Chromatography-Post Column Fluorescence Derivatization (HPLC-FLD), and the fastest way is HPLC-tandem mass spectrometry (HPLC-MS/MS) [67].

HPLC-MS/MS can particularly detect with the greatest level of specificity than the other methods. This method is ideal when testing pesticide residue and can be optimized by using an HPLC-MS/MS with triple instruments in multiple reaction monitoring mode [66].

6. 2,4-Dichlorophenoxyacetic Acid (2,4-D)

2,4-Dichlorophenoxyacetic acid (usually referred to by its abbreviation, 2,4-D) is a common systemic herbicide used in the control of broadleaf weeds. It is one of the most widely used herbicides in the world, and the third most commonly used herbicide in North America [67]. 2,4-D is a synthetic auxin (plant hormone), and as such it is often used in laboratories for plant research and as a supplement in plant cell culture media, such as MS medium. It was a major ingredient in Agent Orange alongside its chemically similar relative, 2,4,5-T (2,4,5-trichlorophenoxyacetic acid).

IUPAC name
(2,4-dichlorophenoxy)acetic-acid
Other names
2,4-D hedonal trinoxol
2',6'-Diethyl-N-butoxymethyl-2-chloroacetanilide [66]

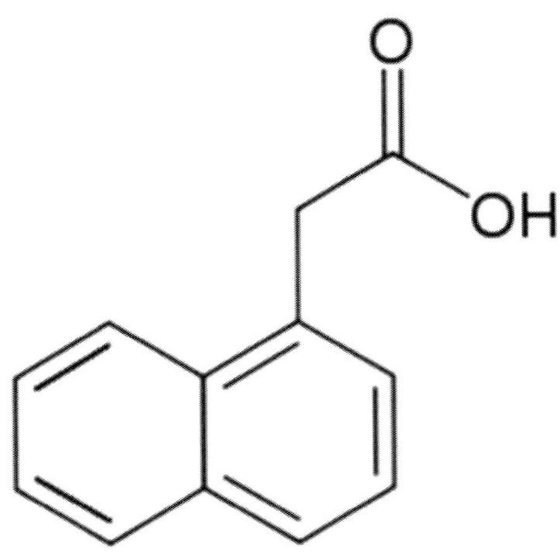

Figure 9. Chemical structure of 1-Naphtalene acetic acid (NAA).

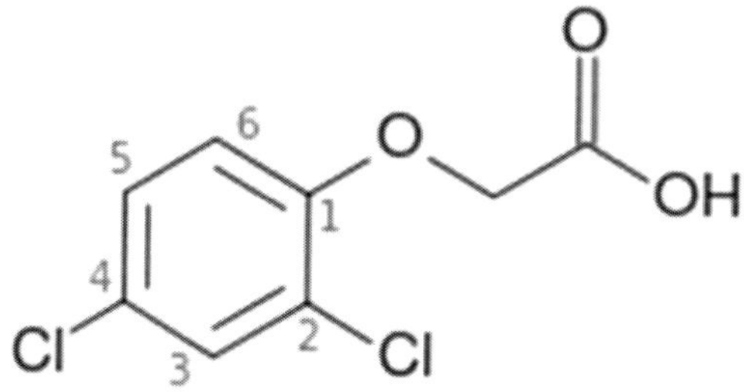

Figure 10. Chemical structure of 2,4-D.

6.1. History

2,4-D was co-discovered independently in both the US and the UK in 1941. The two teams involved were "Templeman and Colleagues at ICI" (US) and "Nutman and Collaborators at Rothamsted Experimental Station" (UK). In both cases the researchers were part of a clandestine wartime effort by their governments, to create chemical warfare agents for use in WWII. For further reading, see section 7.1 of "Herbicides and Plant Physiology By Andrew H. Cobb, John P. H. Reade"

The new chemical's ability to kill weeds was entirely accidental and not the aim of the research in either country. It is important to understand that for war time defense security reasons as well as for international legal reasons, research of this sort was never done openly. Research into chemical warfare agents as well as their production and use were illegal actions under the Geneva Protocol and many other international treaties (see http://en.wikipedia.org/wiki/Geneva_Protocol#Chemical_weapons_prohibitions) which were signed by both the US and the UK. This meant that a plausible cover story was needed to hide the actual nature of any chemical warfare research.

For both the Allies and the Axis powers, the civilian label of agriculture was often used to conceal illegal and/or secret chemical warfare research. Fritz Haber (the Father of Chemical Warfare) used the same method of concealment for the chemical warfare research that he conducted between WWI and WWII, while Germany was still under international sanctions banning them from any form of military research, or the possession of any military forces. Haber's institute did research for the Nazi party that eventually led to the creation of Xyclon, the pesticide that was used as a nerve gas in Auschwitz in the Holocaust. Ironically this invention resulted in the death of nearly all of Haber's relatives. Despite the fact that Haber was a Nobel prize winning Chemist, and as such quite valuable to the German military, he was also of Jewish extraction. Haber eventually had to flee Germany for his own safety, although his institute continued on without him.

Experimentation had become quite common in the fashion of British team at Rothamsted Experimental Station, under the leadership of Judah Hirsch Quastel. Some discoveries were able to be eventually used to increase crop yields, although 2-4D was not used until 1946 near the end of the WWII [70].

When it was commercially released in 1946, it became the first successful selective herbicide and allowed for greatly enhanced weed control in wheat, maize (corn), rice, and similar cereal grass crops, because it kills dicots (broadleaf plants), but not most monocots (grasses).

6.1.1. Genetically Modified Crops

Dow has demonstrated soybean resistance to 2,4-D due to insertion of a bacterial aryloxyalkanoate dioxygenase gene [71]. This is intended as an alternative to Roundup Ready crops due to the increasing prevalence of glyphosate resistant weeds. Department of Agriculture approval is pending for a 2,4-D resistant corn [72].

Unfortunately, in practice previous attempts at the creation of commercial defoliant resistant plant varieties (specifically the "Roundup Ready" varieties of soy and Canola) have a much lower crop yield than expected and also require a larger ongoing investment in agro chemicals by farmers.

"Roundup Ready" varieties have been at the center of much controversy and has been at the center of many protracted legal battles. Acceptance of "Roundup Ready" seed by farmers as of 2013 has been mixed. This may lead to difficulty in launching the 2-4D resistant varieties as well.

6.2. Mode of Action

2,4-D is a synthetic auxin, which is a class of plant hormones. It is absorbed through the leaves and is translocated to the meristems of the plant. Uncontrolled, unsustainable growth ensues, causing stem curl-over, leaf withering, and eventual plant death. 2,4-D is typically applied as an amine salt, but more potent ester versions exist as well.

6.3. Manufacture

2,4-D is a member of the phenoxy family of herbicides, which include:

- 2,4,5-Trichlorophenoxyacetic acid (2,4,5-T)
- 2-Methyl-4-chlorophenoxyacetic acid (MCPA)
- 2-(2-Methyl-4-chlorophenoxy)propionic acids (mecoprop, MCPP)
- 2-(2,4-Dichlorophenoxy)propionic acid (dichlorprop, 2,4-DP)
- (2,4-Dichlorophenoxy)butyric acid (2,4-DB)

2,4-D is manufactured from chloroacetic acid and 2,4-dichlorophenol, which is itself produced by chlorination of phenol. Alternatively, it may be produced by the chlorination of phenoxyacetic acid. The production processes creates several contaminants including isomers, monochlorophenol, and other polychlorophenols and their acids [73].

6.3.1. Dioxin Impurities

Some preparations of 2,4-D are contaminated with dioxins due to the manufacturing process. [72] Tetrachlorodibenzo-p-dioxin (TCDD) is classified as "carcinogenic to humans" by International Agency for Research on Cancer [75].

The typical smell of 2,4-D is the break-down product 2,4-dichlorophenol. 2,4-D is toxic to the liver at small dosages. Increases in liver function tests, jaundice, acute hepatitis, lobular and portal inflammation indicative of a toxic reaction, as well as permanent damage leading to cirrhosis in exposed golfers [76-78]. The defoliant and herbicide Agent Orange, used extensively throughout the Vietnam War, contained 2,4-D. The controversies associated with the use of Agent Orange were associated with a contaminant (dioxin) in the 2,4,5-T component [79].

6.4. Applications

2,4-D is primarily used as a herbicide [80].
2,4-D is most commonly used for:

- Weed control in lawns and other turf
- No-till burndown
- Control of weeds and brush along fences and highway and railroad rights of way
- Conifer release (control of broad-leaf trees in conifer plantings)
- Grass hayfields and pastures
- Cereal grains
- Corn and sorghum (occasionally)
- As a synthetic auxin analog

2,4-D continues to be used, where it is legal, for its low cost. However, where municipal lawn pesticide bylaws exist, such as in Canada, [81] alternatives such as corn gluten meal and vinegar-based products are increasingly being used to combat weeds.

6.5. Toxicity

6.5.1. Cancer Risk

Studies have yielded conflicting results, and different organizations have taken different stances, with regard to the cancer risk of 2,4-D. In 1987, the International Agency for Research on Cancer (IARC) concluded that the phenoxy acid herbicides including 2,4-D, MCPA and 2,4,5-T as a group were classified as a class 2B carcinogen - "possibly carcinogenic to humans" [82]. A 1990 study of farmers in Nebraska, even when adjusting for exposure to other chemicals, found that 2,4-D exposure substantially increased the risk of non-Hodgkin's lymphoma (NHL) [83].

A 1995 panel of 13 scientists reviewing studies on the carcinogenicity of 2,4-D had divided opinions. None of the scientists thought the weight of the evidence indicated that 2,4-D was a “known” or “probable” cause of human cancer.

The predominant opinion indicated that it is possible that 2,4-D can cause cancer in humans, although not all of the panelists believed the possibility was equally likely: one thought the possibility was strong, leaning toward probable, and five thought the possibility was remote, leaning toward unlikely. Two panelists believed it unlikely that 2,4-D can cause cancer in humans [84].

A 2000 study of 1517 former employees of Dow Chemical Company who had been exposed to the chemical in manufacturing or formulating 2,4-D found no significant increase in risk of mortality due to NHL following 2,4-D exposure, but did find an increase in risk of mortality due to amyotrophic lateral sclerosis [85]. On August 8, 2007, the US Environmental Protection Agency issued a ruling that stated that existing data do not support a link between human cancer and 2,4-D exposure [86].

6.5.2. Other

The LD_{50} determined in an acute toxicity rat study is 639 mg/kg [87]. Single oral doses of 5 and 30 mg/kg body weight did not cause any acute toxic effects in human volunteers. This chemical has been associated with the risk of amyotrophic lateral sclerosis [88].

The amine salt formulations can cause eye damage (blindness) on contact; ester formulations are considered non-irritating to the eyes.

One study found that occupational exposure to 2,4-D caused male reproductive problems, including dead and malformed sperm [89].

Concerns regarding neurotoxicity have been voiced with increased sensitivity to amphetamine and thus concerns of increased risk of drug addiction among those exposed [90].

6.5.3. Treatment

While urinary alkalinisation has been used in acute poisonings, evidence to support its use is poor [91].

6.6. Environmental Behaviour

Owing to the longevity and extent of use, 2,4-D is among the most thoroughly studied herbicides with respect to environmental properties. 2,4-D applied at 1.16 lb/acre to bluegrass turf in a laboratory experiment had a half-life of ten days. Other studies found half-life figures between 1.5 and 16 days. Soil microbes are primarily responsible for its disappearance in soil. Studies in Alaska and Canada failed to detect leaching in 22 weeks or from spring to fall, [92] but 2,4-D has been included on the EPA list of compounds that are likely to leach from soil.

In aquatic environments microorganisms readily degrade 2,4-D and breakdown by sunlight is not a major reason for loss. Rates of breakdown increase with increased nutrients, sediment load and dissolved organic carbon. Under oxygenated conditions the half-life can be short, in the order of one week to several weeks. 2,4-D interferes with normal plant growth processes. Uptake of the compound is through leaves, stems and roots; however, it is, in general, nonpersistent. In one study when 2,4-D was applied to grass, there were 80 ppm at day zero, 45 ppm at 14 days, and 6 ppm at 56 days. Breakdown in plants is by a variety of biological and chemical pathways [93].

A number of 2,4-D-degrading bacteria have been isolated and characterized from a variety of environmental habitats [94, 95]. Metabolic pathways for the compound's degradation have been available for many years, and genes encoding 2,4-D catabolism have been identified for several organisms.

As a result of the extensive metadata on environmental behavior, physiology and genetics, 2,4-D was the first herbicide for which the bacteria actively responsible for in situ degradation was demonstrated [96]. This was accomplished using the technique of DNA-based stable isotope probing, which enables a microbial function (activity), such as degrading a chemical, to be linked with the organism's identity without the need to culture the organism involved [97]. This advancement has been particularly beneficial for the study of soil microorganisms, since only a very small fraction of the thousands of species present in highly diverse soil bacterial communities can be isolated in pure culture.

Despite its short half-life in soil and in aquatic environments, the compound has been detected in groundwater supplies in at least five States and in Canada [98]. It has also been detected in surface waters throughout the United States at very low concentrations.

6.7. Legal Issues

2,4-D has been evaluated by the European Union and included on its list of approved herbicides, stating inter alia that "the review [of 2,4-D] has established that the residues arising from the proposed uses, consequent on application consistent with good plant protection practice, have no harmful effects on human or animal health [99]. Concern over 2,4-D is such that it is currently not approved for use on lawns and gardens in Sweden, [100] Denmark, Norway, Kuwait and the Canadian provinces of Québec [101] and Ontario [102]. 2,4-D use is severely restricted in the country of Belize. In 2005, the United States Environmental Protection Agency approved the continued use of 2,4-D [103]. In Canada, the Pest Management Regulatory Agency (PMRA) has placed a condition of registration on 2,4-D such that the 2,4-D registrant(s) must provide the PMRA with a required developmental neurotoxicity study by September 20, 2009 [104].

According to the PMRA, the due date of the study has since been extended to early 2010. On April 18, 2012, EPA denied the petition filed November 6, 2008 by the Natural Resources Defense Council (NRDC) to revoke all tolerances and to cancel all registrations of 2,4-D. EPA stated that recent new study and EPA's comprehensive review confirmed EPA's previous finding that the 2,4-D tolerances are safe [105-106].

7. Effects and Applications of Auxins on Rooting

Adventitious root formation is the key step in successful vegetative propagation and deployment of crop plants. Rooting of cuttings is determined by a complex interplay between internal and environmental factors [107]. Propagation success is thus affected by the age and health status of the stock plant, seasonal timing, propagation technique, and environmental conditions. Among internal factors, plant hormone auxin is the dominant regulator of root development [108].

Although indole-3-acetic acid (IAA) is the most abundant natural auxin, indole-3-butyric acid (IBA) was reported, so far, as the most effective, commonly used auxin to promote adventitious root formation in blueberries and related species propagated from cuttings as well as from tissue culture [109-111]. A favourable rooting effect of IBA may be due to its higher stability in comparison to IAA, its slow transport from the site of application at the base of cutting and its conversion to IAA in the cutting [112-113]. Because of the economic importance in horticulture, agriculture and forestry, development and screening of new rooting compounds potentially useful for the plant production is constantly subject of interest [114].

In several plant species, rooting remains one of the most critical stages in the micropropagation technique, nevertheless, the success of in vitro methods in plant propagation depends not only on the number of plantlets produced but also to their survival rate upon transfer to nursery and field conditions. Several factors such as concentration of auxin, auxin type and concentration affect in-vitro rooting stage [115].

Auxins and their synthetic analogues are commonly used for rooting of cuttings, but their efficiency depends on experimental set-up and, even more importantly, on species or cultivar, and type of explants investigated. In the blueberry cultivar "Bluecrop", 10 g/L IAA or IBA increased the rooting percentage whereas lower concentrations (1.25-5 g/L) had absolutely no effect. In the cultivars Bluetta and Burlington, 2.5 g/L IBA significantly reduced the rooting percentage while IAA, irrespective of concentration did not alter substantially the rooting percentage. On the other hand, both IAA and IBA (2.5-10 g/L) hardly affected the rooting percentage of blueberry plants cv. Jersey [116].

Adding auxin to the medium such as IAA (0, 0.3, 1.5, 3.0, 6.0, or 15.0 mg/L), IBA (0, 0.1, 0.2, 0.5, 1.0, or 3.0 mg/L), and NAA (0, 0.1, 0.2, 0.5, 1.0, or 3.0 mg/L) reduced root length for two winter-hardy roses (*Rosa kordesii* Wulff 'John Franklin' and 'Champlain') and two hybrid teas (*Rosa hybrida* 'John Paul II' and 'Landora') [117].

In general, high level of auxin promoted the production of adventitious roots, although the auxin inhibited the elongation of root [118]. Contrary to our result, in Malus [119] IAA was effective for the production of roots and NAA strongly inhibited the growth of roots. However, other research indicates that IBA increased meristematic activity and advanced the formation of root initials [120], but the difference in rooting between easy and difficult-to-root pear (*Pyrus communis* L.) cultivars was in the uptake of IBA and its conversion to free IAA [121]. Although the known metabolites of NAA [120] are different from those of IBA, their effects on nutrient absorption and their interaction may be similar.

Kim et al. [123] found that IBA treatment was more effective for lateral root induction and root growth compared to NAA in *Panax ginseng*. Hossain et al. [124] found that among the three auxins (IBA, IAA and NAA) tested for root induction of *Zyziphus jujube* Lam., 1 mg/L IBA was found more effective in root production compared to others while the inclusion of NAA (0.5-2 mg/L) in the medium induced low rate of rooting. The superiority of IBA over other auxins regarding *in vitro* rooting is supported by the result of other studies like Sreekumar et al. [125] in *Hemidesmus indicus*, Mala and Bylinsky [126] in *Daphne cneorum*, Kannan et al. [127] in *Caesalpinia bonduc* and Gantait et al. [128] in *Dendrobium chrysotoxum*. The variation in effectiveness of different auxin sources may be attributed to their differential affinity to auxin receptors involved in the rooting process, which may be cultivar specific [129] and depend on the plant type [130].

Rooting of a "difficult – to – root " almond cultivar, "Ferragnes" was induced quikly in the dark in about 55% of the microcuttings using the medium of Bourgin and Nitsch 1967 (BN) containing 0.9% agar and 1 mg /L NAA or 1 mg /L IBA. [131]. Rooting of almond (*Prunus dulcis* batsch) cultivar "Ferraduel" walnut (*Juglans regia* L.) unknown ecotype and chestnut (*Castanea sativa* Miller) occurred with 5 ml Bourgin and Nitsch [132] medium with or without 5 μM NAA and 0.7% agar . Moreover, basal shoot hardening was essential for root formation in chestnut, walnut and almond [133]. Microcuttings of almond genotype (Sel. M51) excised from 15 day old culture 2 to 2.5 cm in length were induced to root on (BN) medium [132], supplemented with IBA (10 μM). The highest rooting percentage was obtained with IAA + light [134]. Moreover, microcuttings of two almond genotypes (cultivar supernova and rootstock M51) treated with darkness (12 days) and two auxins IAA (10 μM) produced rooting in vitro on Bourgin and Nitsch medium [132] with best rooting results of cv. Supernova in the treatment of IBA + darkness (58% rooting percentage).While rootstock M51 showed the best results with IAA + light (95.0% rooting percentage) [135]. Almond seedlings (*Prunus dulcis* Mill) Sel. M51 rooted on MS medium [136] supplemented with IAA (5 μM) [137]. The main role of auxin in the culture is concerned in root initiation, which occurs by stimulation of cell division, which results by auxin promotion of cambial activity [138].The addition of NAA in the range of 0.1 – 2.0 mg/L to dilute modified MS medium was recommended in rooting of proliferated shoots of many prunes spp. [139]. Rooting of almond cultivar "Ferragnes" was induced quickly in the dark in about 55% of microcutting using Bourgin and Nitsch (BN) [132] with only the macronutrients reduced to half, but containing 0.9 % agar and 1 mg / L NAA or 1 mg / L IBA. [131]. Microcuttings of almond genotype (Sel.M51) were treated with darkness (12 days) and IBA or IAA both at 10 μM to induce rooting in vitro. The best rooting results (95.0%) were obtained with IAA and light on BN medium. [134]. Microcuttings of seven genotypes of almond (*Prunus dulcis* Mill) were excised from 15 – d – old cultures and transferred into rooting medium consisting of MS salts supplemented with IBA at (5 μM) . Best rooting percentage 75.0% was induced with almond rootstock M51 by IBA treatment at (5 μM). [137]. Using either 2.0 or 4.0 mg/L concentrations of auxin (IBA, NAA) encouraged root formation of almond, peach and apricot plants [140]. The highest rooting capacity (90.0%) was obtained without callus formation in the medium supplemented with 4.9 μM IAA after 5days in the light of M51 almond rootstock [141]. The best rooting percentage of Prunus nume Sieb .Et Zucc. was obtained on WPM medium supplemented with 1 μM NAA [142]. NAA was the most effective auxin in inducing the best rooting percentage (40.74), greatest number of roots/rooted shoot and the longest root of bitter almond shoots followed by IBA, while IAA took the other way around [143].

The best-developed root system and the highest mass were obtained from MS medium with 1.0 mg/dm^{-3} IBA, whereas the longest roots mean for the strawberry cultivars 'Elsanta' and 'Senga Sengana' on MS with 0.5 and 1.0 mg/dm^{-3} NAA [144]. In the apple rootstock Merton I. 793, low concentrations of α-NAA was proved to be more effective for rooting, as compared to IBA and IAA [145]. Indole-3-butyric acid (IBA) was used in rooting of shoots of *Bixia orellana* [146], *Digital lanata* [147], *Picrorhiza kurroa* [148], *Bacopa* [149] and *Columnea* [150-152]. While indole-3-acetic acid (IAA) was used in rooting of microshoots of *Ocimum gratissimum* [152] and *Nemathanthus hybridus* [153].

The combination 0.5 mg/L NAA + 1.0 mg/L IBA resulted in the greatest rooting percentage and mean root length in half-strength MS solid medium of *in vitro*-derived leaf and hypocotyl explants of Chicory (C*ichorium intybus* L. cv. Focus), under total darkness.

On the other hand, the combination 0.5 mg/L NAA + 0.5 mg/L IBA proved to be the best as regards the mean number of roots [130]. The combination 0.2 mg/L NAA + 0.5 mg/L IBA resulted in the greatest rooting percentage and mean root fresh weight in half-strength MS liquid medium of *in vitro*-derived leaf and hypocotyl explants of Chicory (C*ichorium intybus* L. cv. Focus), under total darkness [130]. In Malus, IAA was effective for the production of roots and NAA strongly inhibited the growth of roots [119].

In *Paulownia kowakamii,* addition of 1.0 mg/L IAA to half strength MS medium gave 100% rooting percentage, however , the highest number of roots formed per shootlet (5.53) were observed with culture media contained (1.0 mg/L) NAA [154]. The highest rooting of *Prunus serotina* Ehrh. shoots and number of roots per shoot was obtained with 2.5 μM IBA

[155]. Of two different auxins tested for *in vitro* rooting, IBA was most effective compared to NAA. Half-strength MS medium containing IBA at an optimum concentration of 2.0 mg/L induced rooting in 83.1% of the *in vitro* derived shoots of *Withania somnifera* (Ashwagandha) [156]. In shoot tips explants of *Rosa gruss an teplitz* and *Rosa centifolia,* IBA increased rooting percentage (89.38), number of roots (8.72) and root length (3.58 cm) more efficiently at 0.50 mg/L [157]. Nowadays IBA is used commercially to root microcuttings and is more efficient than IAA [158].

Among the indolebutyric acid (IBA) and naphthaleneacetic acid (NAA) hormones, tested for maximum percentage of rooted shoots, root number, root length and quality of roots, IBA at 1.5 mg/L concentration proved to be the best one for rooting of Moraiolo cultivar of olive producing maximum root initiation in 86.67% shoots, 5.03 number of roots per rooted explant and 4.95 cm root length. The roots produced on IBA were longer with better quality shoots whereas NAA produced poor response with necrotic leaves and leaf abscission [159]. Benelli et al., [160] and Tanimoto [161] have proved that IBA is the most effective auxin in olive rhizogenesis, as compared to NAA. The reason for these differences in root inducing ability may be the slow and continuous release of IAA from IBA [162-163] and release of IBA through hydrolysis of conjugates [158]. These IBA conjugates were reported to be superior to free IBA in serving as an auxin source during later stages of rooting [164]. The inferior effect of NAA on the root number may be due to the reason that NAA is more persistent than IBA, remains present in the tissue and may block further development of root meristemoids [119, 165]. Hausman [166] has shown that in tissue culture media, IAA is photo-oxidized rapidly (50 % in 24 h) while IBA oxidized slowly (10 %) and NAA is very stable.

In *date palm Barhi cv.,* the highest root number was obtained when the plants treated with IBA compared with NAA and the best concentration of IBA was 10 mg/L. The best results for root length were obtained when the date palm plants were treated with 10.0 mg/L IBA for 18 months (64.4 cm) [167]. IBA was less effective than IAA or NAA in stimulating root production on sainfoin cuttings (*Onobrychis viciifolia*) [168]. Various concentrations of IBA initiated rooting but maximum rooting percentage was found with 2 and 2.5 mg/L IBA in MM106 [169]. In carnation (*Dianthus caryophyllus* L.), the highest root percentage for Eskimo Mogr cultivar (83%) was achieved on the medium supplemented with 0.5 mg/L NAA, while for Innove Orange Bogr cultivar (98%) it was obtained on the medium containing 1.5 mg/L IBA [170]. In *Amygdalus communis* cv. Nonpareil shoots cultured on a half strength MS medium supplemented with 8.0 mg/L IAA showed both primary and secondary root formation at the rate of 50% [171]. In *Thymus vulgaris* L. plantlets, the best root apparatus (92.5% rooting rate, with 19 adventitious roots per shoot) developed on MS medium supplemented with 0.05 mg/L 2,4-dichlorophenoxyacetic acid [172].

Rooting of the mature clone of *Acacia mangium* was significantly increased by exposing the microshoots to auxins (4 and 6 μM IAA or IBA) in darkness, while the promoting effect of darkness combined with 4 μM IAA was more time-restricted for the juvenile-origin microshoots [173]. One hundred percent of cloned *Citrus limon* L. shoots rooted *in vitro* on half-strength MS medium +27 μM NAA +0.1% activated charcoal [174]. The best rooting results in *Metrosideros excelsa* Sol ex Gaertn. "Parnel"were accomplished with IAA at 5.71 μM (89% rooting) and with IBA at 2.85 or 5.71 μM [175]. Although Nobrc et al. [176] found 1.3 μM NAA to be the optimum concentration to induce rooting of another Viburnum species (V. tinus), NAA concentration above 1.5 μM caused formation of callus at the base of the microcuttings and tufted roots in sweet viburnum [177].The rooting of Vitis thunbergii Sieb.

et Zucc. shoot tips was optimized using MS medium supplemented with 0.5 mg/L napthaleneacetic (NAA) with eight roots, 3 cm long after 1 month of culture [178]. However, a rooting medium with a combination of 2.5 μM IBA and 2.5 μM NAA was most suitable for *in vitro* rooting of *Eucalyptus sideroxylon* explants [179].

8. Effect of Auxins on Chlorophyll, Carotenoid, Carbohydrate and Proline Content

Carotenoids are polyen pigments that widely occur in animal and plant kingdom in the free form or as esters of fatty acids. Carotenoids are indispensable components of photosynthetic pigment-protein complexes. They serve as light-harvesting pigments and play an important role in protecting the photosynthetic apparatus against the damaging effects of singlet state oxygen. Singlet state oxygen in the photosynthetic apparatus can be formed by energy transfer from triplet-state chlorophyll to ground-state oxygen [180-184].

Carotenoids pigments, along with chlorophylls, play an important role in photosynthesis, especially in algae existing in deeper water strata. They perform a role of direct light energy absorbents. Carotenoids prevent chlorophyll from photooxidation and perform essential role as a constituent element of photoreceptors and sexual chemotaxis of algae mobile cells. They also perform a very important role in oxidation-reduction connected with oxygen, hydrogen and electron transport. Some of carotenoids are precursors of plant hormones, for example abscisic acid (ABA) [185-186].

IAA exerts influence on plant growth by enlarging leaves and increasing photosynthetic activities in plants [187]. IAA also activates the translocation of carbohydrates during their synthesis [188-189]. The total carbohydrate content of the tomatoes (*Solanum lycopersicon*), genotype JM 94/47treated with the IAA, IBA and NAA were statistically lower as compared to the control [190].

This suggests that application of these bioregulators suppress the photosynthetic process through their influence on certain enzymatic action [191]. The suppression might be responsible for the observed decrease in carbohydrate content [192]. Total chlorophyll content of *Jatropha curcas* L. leaves was found to be stimulated significantly by the spray of IAA and NAA [193]. Chlorophyll and protein loss in leaves are conventionally used biochemical parameters to indicate senescence [194]. Leaf senescence is marked by degradation in chlorophyll levels.

In detached leaves of *Tropaeolum majus* L., a decline in chlorophyll and carotenoid content were evident in the leaves of both control and treated plants after their transfer to dark, but the extent of the decline in chlorophyll content in treated plants was significantly higher than in control plants [195]. Our results are convenient to the previous study [196] in which they showed that NAA treatment in 14 week-aged tobacco plants increased the loss of chlorophyll. Moreover, high concentration levels of IAA in leaf discs of lettuce also accelerated the chlorophyll loss [197]. However, IAA was suggested to prevent the loss of chlorophylls throughout the aging of chloroplasts in wheat both *in vivo* and *in vitro* conditions [198]. Plant tissues contain substantial amounts of carotenoids that serve as scavenger compounds to oxygen radicals [199].

It was suggested that IAA has retarded chlorophyll loss from leaf discs of romaine lettuce but accelerated chlorophyll loss in a high concentration of IAA (10^{-4} M) [197]. However, IAA was suggested to prevent the loss of chlorophylls throughout the aging of chloroplasts in wheat both *in vivo* and *in vitro* conditions [198]. Many studies have suggested the correlation between auxin levels with senescence and abscission [200]. One of them was reported that a gradient of auxin levels between the leaf blade and the stalk occurred in bean leaves. Auxin levels declined as leaf aged and senescence occurred when auxin levels between the leaf and stalk were approximately equal [201]. Recently, auxin was reported to repress transcription of some genes whose expression is correlated with senescence and/or abscission [202]. According to the literary data, auxin may retard [201] or accelerate [197] senescence in some cases in plants.

Carbohydrate changes are of particular importance because of their direct relationship with such physiological processes as photosynthesis, translocation, and respiration [203]. In *Wolffia Arrhiza* (L.) Wimm. (Lemnaceae), IAA (indole-3-acetic acid) and SA (salicylic acid) strongly stimulated action on the content of chlorophylls a and b and carotenoids (especially β-carotene and lutein + zeaxanthin). On the other hand, the chemical analogue of IAA, i.e. NAA had a generally inhibitory effect on chlorophyll and carotenoid contents [204]. Similar results of increased chlorophyll content with IAA treatment have been reported by Shobha et al. [205] on *Euphorbia antisyphilitica*. Carbohydrate content of cowpea (*Vigna unguiculata* (L.) Walp) Var. Bari Fellon-1 seeds were decreased with the increasing concentration of IAA [206]. In spinach (*Spinacia oleracea* L.) chlorophyll content was not affected by the application of NAA [207]. IBA was correlated with significant increase in total chlorophylls in leaves, total soluble carbohydrates and sugars in seeds of pea (*Pisum sativum* L.) [208].

One of the most important mechanisms exerted by higher plants under environmental-stress conditions is the accumulation of compatible solutes such as proline. The environmental stress induces an increase in proline concentration in plants. Proline accumulation under environmental stress may contribute to osmotic adjustment, protecting cell structure and function or may serve as metabolic or energetic reserve in plants [209-210]. The hormones IAA, IBA and 2-4-D were applied individually in 10 μM and 100 μM concentrations and markedly decreased the proline content in *Bambusa arundinacea* [211]. The application of IAA and NAA enhanced the proline content of the treated *Jatropha curcas* plants compared to the untreated plants [193].

In coleoptiles of wheat seedlings or in excised coleoptile segments incubated under shaking for a 24 h period, the application of high IAA concentrations (more than 10^{-5} M) increased proline accumulation in the presence of 10 mM acetic acid in the pH range from 6.0 to 7.5, in which acetic acid alone was less effective [212].

IBA was correlated with significant increase in proline in seeds of pea (*Pisum sativum* L.) [208]. In *Bambusa arundinacea* seedlings, the hormones IAA, IBA and 2,4-D applied individually in 10μM and 100μM concentrations resulted in a decrease of the proline content. One of the most important responses of plants to drought and other abiotic stresses is the overproduction of different types of compatible solutes [213]. In response to osmotic stress, many plant species accumulate proline due to the simultaneous ABA-mediated activation of its biosynthesis and inactivation of its degradation pathways during stress [214]. Application of NAA showed no effect on proline content at a concentration of 10^{-5} M or a strong decrease in proline content in the 4-days old tomato seedlings (*Lycopersicum esculentum*) compared to control at 10^{-4} M [215]. When IBA was supplied to barley (*Hordeum vulgare*) leaf sections a similar rapid and marked decrease in the intracellular pH was observed, and this acidification was accompanied by an increase in proline level [216].

9. Shoot-Derived Auxin Contributes to Root Organization

Auxins, a class of essential plant hormones, orchestrate almost every aspect of plant growth and development [217-218]. In roots, the most well-characterized auxin-associated phenotypes are the dose-dependent increase in the length of epidermal-derived root hairs, the bimodal effect of auxin concentration on primary root length, the dose-dependent increase in number of lateral root primordia, and the response to gravity [219-222]. Proper output of auxin signaling depends on the interplay of the transport, biosynthesis, conjugation, and signaling of this class of small molecules. Because the role of auxin during embryogenesis is covered in another monograph of this series [223], we focus here on recent findings related to auxin and postembryonic root development. Auxin can be synthesized in young leaves and cotyledons [224]. The same long distance pathway that carries carbohydrates from "source" to "sink" also facilitates the bulk flow of auxin and other hormones such as ABA and cytokinins as well as mRNA and proteins [225]. Because bulk flow provides the driving force through membraneless phloem channels, the rate of auxin movement is rapid, reaching speeds of up to 7 cm/h in the roots of *Populus tremula* and *Vicia faba* [226]. The symplastic isolation of the sieve elements and companions cells along this transport route likely drives the movement of IAA towards the root tip, which represents the major sink tissue. Although auxin is transported over long distances, our understanding of its role during development comes primarily from studies focused on sink tissues such as root tips, where cells are largely undifferentiated and where auxin movement through the apoplast would not be impeded by secondary cell walls. Similarly, the protonation of auxin in the acidic environment of the cell wall facilitates its movement by diffusion across cell membranes. It is not until the molecule becomes ionized in the less acidic cytoplasm that movement through sink tissue becomes limited and requires the involvement of integral membrane transport proteins to control directional movement. Commonly referred to as "polar auxin transport" (PAT) this mechanism of directing auxin distribution clearly differs from the long-distance transport described earlier.

The rates of PAT in roots of *Phaseolus vulgaris, Pisum sativum*, and *Arabidopsis thaliana* are 10 mm/h, which is at least significantly slower than auxin movement through the phloem.

PAT is the outcome of the coordinated activities of several membrane transport protein families that facilitate auxin influx (e.g., AUXIN RESISTANT1 [AUX1] and LIKE-AUX1 [LAX]) [227] and efflux (e.g., members of the P-GLYCOPROTEIN ABC transporter family [228] and the PIN-FORMED [PIN] proteins [229]. The molecular, cellular, and biochemical aspects of these proteins have been the focus of a number of excellent reviews [230-231, 218].

10. Production of Auxins

Indole acetic acid is one of the most physiologically active auxin and a common product of L-Tryptophan metabolism by several microorganisms inducing plant growth-promoting bacteria (PGPR) [232-235]. Rhizobia are known to produce significant levels of IAA both in free living conditions and also symbiotically in nodules [236]. Among the auxins, IAA (indole-3-acetic acid) is recognized as a key factor, which is directly beneficial for plants. IAA is unique in being implicated both directly in plant growth promotion/formation of symbiotic associations, as well as indirectly in pathogenesis; therefore its contribution to plant related physiological processes and microbiological activity in soil is highly complex [237]. The ability to form plant hormones is believed to be a major property of several rhizospheric, epiphytic and symbiotic microorganisms including cyanobacteria, besides members of the plant kingdom [238]. Auxins represent a wide group of the compounds which are derivatives of the indole ring. This chemical structure explains the relationship between the quantity and the activity of their members. Since the discovering of auxins many efforts of scientists have focused on the possibility of manipulating plant growth and development. It was suggested that mycorrhizal fungi produce auxins (indole-3-acetic acid – IAA is preferred) and that these auxins affect the growth of susceptible roots by modifying apices so that they assume the structure seen in the host tissues of mycorrhizas [239].

11. Action of Auxins

Auxins are a group of tryptophan-derived signals, which are involved in most aspects of plant development [240]. Auxin plays a major role in controlling growth and development of plants, early stages of embryogenesis, organization of apical meristem (phyllotaxy) and branching of the plant aerial parts (apical dominance), formation of main root, lateral and adventitious root initiation [241]. Auxin is also involved in gravitropism and phototropism [242]. These multiple effects across the plant result from its control of cell division, cell elongation and certain stages of differentiation [243]. Auxin is synthesized mainly in young leaves and is actively transported to other tissues to coordinate growth and facilitate responses to environmental variations.

Indole-3-acetic acid (IAA) has very important roles in plant sexual reproduction: controlling the development of stamens, gynoecia, and ovary, promoting the maturation of egg cells, inducing the axial polarity and polar development of the embryo [243-247]. Some researches also reported the increase of free IAA in the pistil after pollination and suggested that IAA probably promotes pollen tube growth in the pistils [244-245].

A previous study also confirmed that free IAA was present throughout the style tissue and its content increased after pollination in *Torenia fournieri* L. [248]. Interestingly, auxin is able to repress stomatal closure through its antagonistic effect on ABA [249].

Auxin, known to be involved in cell enlargement, was long thought to be the controlling factor in the rooting process. Two types of evidence support this reasoning: (a) the increased content of endogenous auxin in the base of cutting during rooting induction [250] and (b) the rooting response of many plants to exogenous auxin. Sugiyama [251] reported that the effect of an auxin on rooting is promotory at low concentrations and inhibitory at supra-optimal concentrations. The enzymes involved in cell enlargement processes are triggered by the auxin. Wada et al., [252] reported that IBA promotes root length by influencing the synthesis of enzymes concerned in cell enlargement. Ludwig-Muller [253] reported that the superior effects of IBA on root elongation as compared to NAA might be due to the several factors, such as its preferential uptake, transport, metabolism and subsequent gene activation. Taiz and Zeiger [254] reported that roots may require a less concentration of auxin to grow, but root growth is strongly inhibited by its higher level because at this level, auxin induces the production of ethylene, a root growth inhibitor.

It is generally accepted that auxins play a central role in the process of root formation [255, 114]. They induce root initials [256-257] and influence the growth of the newly formed roots in the expressive phase of root development [258]. Many investigations have shown that IBA (indole-3-butyric acid) has a greater ability to promote adventitious root formation in comparison to IAA [114, 253]. It is more stable and less sensitive to the auxin degrading enzymes [256, 158, 260]. IAA is rapidly metabolized by the peroxidase, acting as an IAA-oxidase, with the strongest activity during the root initiation phase [257, 260]. Biochemical studies in numerous plants and genetic studies of *Arabidopsis* IBA-mutants indicate that IBA acts primarily via its conversion to IAA, which occurs through a mechanism similar to peroxisomal fatty acid b-oxidation [256, 262]; however, some evidence suggests that IBA acts as an auxin on its own [256, 262, 253]. Oxidation products of IAA may promote root formation, especially when linked to the phenolic compounds present [263]. Exogenously applied IAA and IBA change the internal auxin level and the level of their conjugates, which positively influence the adventitious rooting of cuttings [264, 256, 260].

The development of callus from immature leaf explants is directly related to the presence of 2,4-D which is a suitable growth hormone responsible for callus induction in most plant species in plant tissue culture work. This is similar to the findings of Mamun et al. [265] and Baskaran et al. [266].

12. Physical and Chemical Properties of Auxins

Natural and synthetic auxins are low-molecular weight, organic substances, containing either an indole or an aromatic ring. They are crystalline and only slightly soluble in water, but readily soluble in organic solvents (ethanol, methanol, acetone, diethyl ether and dimethyl sulphoxide, DMSO) or, as weak acids, in alkaline watery solutions. With the exception of IAA they are stable and persist in the media for tissue cultures. IAA is less stable and is especially sensitive to the light (mainly UV) and to oxidants.

Both naturally occurring indolic auxins and their synthetic counterparts, in spite of their different structures, have similar physiological effects. For auxin activity, an aromatic ring is needed at a distance of 0.55 Å from a carboxyl group. Indeed, all active auxins are weak organic acids. The relative degree of activity of individual auxins in different growth processes is very variable.

It differs not only from plant to plant, but also from organ to organ, tissue to tissue, cell to cell and, moreover, also with the age and physiological state of the plant (tissue) [267]. Probably due to its high instability, IAA is usually less effective than synthetic auxins like 2,4-D or NAA.

13. Metabolism of Auxin

In general, the metabolism of auxin, as well as the metabolism of any other hormone, consists of both biosynthetic and hormone-molecule modifying reactions. Auxin biosynthesis is usually thought to be more intensive in meristematic regions and young growing organs such as rapidly growing leaves, apical buds, root tips, and developing inflorescences.

13.1. Natural Levels of IAA

In intact plants, the highest level of IAA occurs predominantly in or near the apex and IAA is transported basipetally; thus, we can observe a decreasing IAA concentration from the top to the bottom. The above is especially true for monocotyledons, whereas in dicotyledons the highest concentration of IAA occurs in the subapical zone, which also grows most rapidly [268]. Levels of naturally occurring auxin in explanted tissues are found to depend on the mother plant from which the explants are taken.

The age of the mother plant, the conditions under which it has been growing and the season of the year at which explants are taken, can all be influential [269]. As stated above, the time of the day may also be important. The presence of an associated organised meristem can cause callus to grow more vigorously, or to be capable of organogenesis [270]. This suggests that meristematic cells are particularly active sites for the biosynthesis and/or the release of natural growth factors favouring cell growth [271].

13.2. Auxin Biosynthesis

The most common IAA-biosynthetic pathway in plants appears to be the indolepyruvate one; it begins by the transamination of Trp catalysed by the enzyme tryptophan transaminase (or tryptophan amino transferase) [272]. Indolepyruvate is then converted into indole-3-acetaldehyde [273]. IAA can originate from this aldehyde either by dehydrogenation (NAD-dependent indoleacetaldehyde dehydrogenase) or by oxidation (indoleacetaldehyde oxidase) [274]. The indole-3-acetaldoxime pathway starts by formation of indole-3-acetaldoxime followed by its conversion to indole-3-acetonitrile and final hydrolysis (nitrilase) to IAA [275].

13.3. Auxin Conjugation and Degradation

Auxin is known to form two main classes of conjugates: amides (peptides) and glycosyl esters. Amides originate *via* formation of peptide bonds between auxins (both native and synthetic) and amino acids, namely aspartate, glutamate, alanine, glycine and valine. Glycosyl esters of IAA involve compounds such as various isomers of *o*-indole-3-acetyl-β-D-glucopyranose and indole-3-acetyl-*myo*inositol, indole-3-acetylglucosyl-rhamnose, etc. [276]. Some of the auxin conjugates may also serve as transport forms of auxin.

In contrast to conjugation reactions, auxin degradation is an irreversible process. There are two main ways of IAA degradation: oxidative decarboxylation and oxidation without decarboxylation. The oxidative decarboxylation may be catalysed by a non-specific activity traditionally called "IAAoxidase/peroxidase" and leads to the subsequent formation of degradation products such as indole-3-methanol, indole-3-aldehyde, methylene oxindole and indole-3-carboxylic acid [277].

Oxidation without decarboxylation involves complex pathways resulting in the formation of oxindole-3-acetic and dioxindole-3-acetic acids and their derivatives. Similar pathways are functional also for degradation of some conjugates, *e.g.*, indole-3-aspartate (Slovin et al., 1999). The synthetic auxins, 2,4-D and NAA are often converted, after uptake into plant tissues, to conjugates, mainly glucosyl esters [278].

14. Transport of Auxins

Auxins seem to be the only group of plant hormones exhibiting - on the level of the whole plant or its parts - active transport in a polar manner in addition to long-distance movement via vascular tissues [279]. Auxins appear to be transported long distances extensively through the phloem (but probably not the xylem) of higher plants. The phenomenon of polar auxin transport was demonstrated first in the late 1920s [280]. In contrast to the movement of auxin *via* vascular tissues, polar cell-to-cell auxin transport is localised predominantly in parenchyma cells in the sheath surrounding differentiated vascular tissue.

The polarity of auxin transport was explained [281] by the different permeability of opposite parts of cells for dissociated and undissociated molecules of indole-3-acetic acid (IAA^- and IAAH, respectively), and by an asymmetric localisation (basal in the stem cells) of a so-called auxin efflux carrier (translocating the IAA anion outside the cell). This idea was summarised by Goldsmith [282] and named the "chemiosmotic polar diffusion theory".Now, it is believed - on the basis of biochemical, physiological and molecular biological data - that at the level of the individual cell, both passive diffusion and an auxin-uptake (influx) carrier drive auxin translocation into the cell while an auxin efflux carrier drives auxin transport out of the cell. The activity of both carriers, to different extents, can be inhibited by several synthetic compounds, mainly of the phytotropin type, *e.g.* 1-naphthylphthalamic acid (NPA) [283].

15. Physiological Effects of Auxins

Auxin is in most cases active in the concentration range 0.1 – 10 μM. Activation of cell elongation by auxin is mediated by increased proton efflux ('acid growth theory') [284] and/ or changes in gene expression. In nature, an increase in cell division is most obvious in the spring in trees, when young buds produce auxin, which stimulates cell division in the cambium [285].

Auxins stimulate differentiation of vascular bundles and, as already discussed, they take part in differentiation of buds and roots [286]. Auxin is gradually canalised by a positive feedback mechanism where increasing conductivity of auxin conducting cells leads to canals of cells efficiently transporting auxin [287]. Polar transport of auxin is fundamental for the establishment and maintenance of polarity of the plant and its organs. Inhibition of polar auxin transport leads to many abnormalities and in embryos it can lead to death [288].

Auxins are known for their ability to promote adventitious root formation. IBA is by far the most commonly used auxin to obtain root initiation in conventional cuttings. It has been shown that IBA is readily converted to IAA, but it probably also has an effect on its own [289].

Polar transport of auxin is the decisive force of apical dominance [290]. Removal of the tip, the main auxin source, or inhibition of auxin transport leads to the outgrowth of axillary buds. Also dominance of fruits is mediated by auxin transport. Uneven distribution of auxin is considered to cause differential growth rates in different sides (upper/lower or irradiated/ shaded) of coleoptile or root and their bending in gravitropic or phototropic reactions [291].

16. Auxin Effects in Tissue Culture

Some reduction of chlorophyll formation in the presence of 2,4-D was noted in callus cultures of pea, tomato and potato by Hildebrandt et al., [292] and *Oxalis dispar* callus was found to turn green only when the auxin content in the medium was reduced to one tenth of that normally used to promote callus growth (*i.e.*, 2,4-D reduced from 5 to 0.5 μM, or NAA from 50 to 5 μM). Also, development of lateral roots is stimulated by auxin as was demonstrated in *Panax ginseng*, where IBA was shown to be more effective than NAA [293]. In alfalfa, low 2,4-D concentration gives rise to callus from leaf explants, while higher 2,4-D levels induce formation of embryo-like structures [294]. In the induction of somatic embryogenesis from immature cotyledons of *Glycine max*, Lazzeri et al., [295] discovered a highly significant interaction between the concentration of auxin and sucrose in the medium. In the study of somatic embryogenesis in *Phaseolus vulgaris* induced by 2,4-D, it was shown that there was an inverse relationship between 2,4-D concentration in the medium and endogenous IAA level in the globular structures [296]. Using an immunocytochemical approach to visualize IAA it was shown in sunflower somatic embryos that an endogenous auxin pulse induced by the application of exogenous auxin is needed as one of the first signals inducing embryogenesis [297].

Heyser et al., [298] found that in *Triticum aestivum*, some varieties produced embryogenic callus with 2,4-D. At high concentrations most synthetic auxins are phytotoxic to field-grown broad-leafed plants. 2,4-D has been used commercially as selective herbicide.

This detrimental effect was reported also for high levels of IAA and is probably due to a dramatic increase in ethylene production. A mixture of 2,4-D (or 2,4,5-T) and either IAA or L-tryptophan was found to promote embryogenic callus formation in wheat, pearl millet and some varieties of rice [299]. This observation was confirmed by Carman et al., [300], who found that adding 0.2 mM L-tryptophan to 3.62 μM 2,4-D, consistently enhanced the formation of somatic embryos from scutellum callus of wheat genotypes. Mixtures of auxins are also more effective in inducing regeneration of wheat, barley and triticale [301].

IAA and synthetic auxins such as NAA and 2,4-D are rapidly taken up into cultured tissues from media with a pH less than 5-6. The compounds are subsequently absorbed into cells as whole molecules (via uptake carrier or diffusion, see above), but dissociation then causes them to be retained within the cell, because the plasmalemma is impermeable to auxin anions [302]. In apple microcuttings, applied auxin is taken up predominantly via the cut surface and not via the epidermis [303]. Total 2,4-D uptake (14C-labelled) was found to be higher in easy-to-root juvenile clones of *Sequoiadendron giganteum* than in difficult-to-root mature stem cuttings [304]. Another comparison of 2,4-D uptake, distribution and metabolism was performed with explants of cucumber (*Cucumis sativus* L.) hypocotyls and cotyledons. Cotyledon explants take up more 2,4-D, have a more pronounced basipetal gradient of 2,4-D level and conjugate 2,4-D more actively than hypocotyl explants [305].

In the cherry rootstock CAB-6P (*Prunus cerasus* L.), IAA and IBA at 1 mg/L resulted in an increased number of roots per rooted explant but in a decreased root length. In CAB-6P (*Prunus cerasus* L.) and Gisela 6 (*Prunus cerasus* L. x *Prunus canescens* L.) explants, IBA was proved to be the best auxin regarding the number of roots and rooting percentage whereas IBA gave the shortest roots. In MxM 60 (*Prunus avium* L. x *Prunus mahaleb* L.) cherry rootstock, IBA gave better rooting results than IAA concerning the number of roots per rooted explant and rooting percentage whereas the root length was greatest in a medium free of auxins [306].

Conclusion

Auxins are an important category of acetic acids influencing a number of physiological and developmental processes in plants. The findings indicate that the four auxins i.e. IAA, IBA, NAA and 2,4-D have a direct effect on *in vitro* rooting, are involved in the photosythetic apparatus, influencing leaf chlorophyll content and participate on carbohydrate biosynthesis and metabolism, as well as in proline accumulation in both leaves and roots. In tissue culture media, IAA is photo-oxidized rapidly (50 % in 24 h) while IBA oxidized slowly (10 %) and NAA is very stable. The effectiveness of the four auxins on the rooting process is concentration and type dependent and it is different among the various plant species.

References

[1] IUPAC, Commission on Nomenclature of Organic Chemistry. Table 28(a) Carboxylic acids and related groups. Unsubstituted parent structures. *A Guide to IUPAC Nomenclature of Organic Compounds*. Blackwell Scientific publications; 1993.
[2] Acetic Acid - PubChem Public Chemical Database. *The PubChem Project*. US: National Center for Biotechnology Information.
[3] IUPAC Provisional Recommendations 2004 Chapter P-12.1; page 4
[4] *Scientific literature reviews on generally recognized as safe (GRAS) food ingredients*. National Technical Information Service. 1974. p. 1.
[5] Chemistry. Encyclopedia Britannica, 1961 5, 374.
[6] UK Food Standards Agency: Current EU approved additives and their E Numbers. Retrieved 2011-10-27.

[7] US Food and Drug Administration: Listing of Food Additives Status Part I. Retrieved 2011-10-27.

[8] Australia New Zealand Food Standards Code"Standard 1.2.4 - Labelling of ingredients". Retrieved 2011-10-27.

[9] Cheung, H.; Tanke, R. S.; G. Torrence, P. *Acetic Acid in Ullmann's Encyclopedia of Industrial Chemistry.* Weinheim: Wiley-VCH; 2005.

[10] IUPAC Provisional Recommendations 2004 Chapter P-12.1; page 4

[11] Armarego, W. L. F.; Chai, C. *Purification of Laboratory Chemicals,* 6th edition. Butterworth-Heinemann; 2009.

[12] Geoffrey, M. *Industrial and Manufacturing Chemistry* (Part 1, Organic ed.). London: Crosby Lockwood; 1917; 330-331.

[13] Adet, P. A. Mémoire sur l'acide acétique (Memoir on acetic acid). *Annales de Chemie,* 1798 27, 299-319.

[14] Goldwhite, H. Short summary of the career of the German organic chemist, Hermann Kolbe. *New Haven Section Bull. Am. Chem. Soc.,* 2003 20.

[15] Schweppe, H. Identification of dyes on old textiles. *J. Am. Inst. Conservation,* 1979 19, 14-23.

[16] Wagner, F. S. Acetic acid. In: Grayson, M., editor. *Kirk-Othmer Encyclopedia of Chemical Technology.* 3rd edition. New York: John Wiley and Sons; 1978.

[17] Lancaster, M. *Green Chemistry, an Introductory Text.* Cambridge: Royal Society of Chemistry; 2002; 262-266.

[18] Mehringer, D. M.; Snyder, L. E.; Miao, Y.; Lovas, F. J. Detection and confirmation of interstellar acetic acid. *Astrophysical Journal Letters,* 1997 480 L71-L-74.

[19] Jones, R. E.; Templeton, D. H. The crystal structure of acetic acid. *Acta Crystallogr.,* 1958 11, 484-487.

[20] Briggs, J. M.; Nguyen, T. B.; Jorgensen, W. L. Monte Carlo simulations of liquid acetic acid and methyl acetate with the OPLS potential functions. *J. Phys. Chem.,* 1991 95, 3315-3322.

[21] Togeas, J. B. Acetic acid vapor: 2. A statistical mechanical critique of vapor density experiments. *J. Phys. Chem. A,* 2005 109, 5438-5444.

[22] Zieborak, K.; Olszewski, K. *Bull. Acad. Pol. Sci. Ser. Sci. Chim. Geol. Geogr.,* 1958 6, 3315-3322.

[23] Buckingham, J. *Dictionary of Organic Compounds.* 6th edition. London: Chapman and Hall. 1996.

[24] Production report. *Chem. Eng. News.* 11 July 2005 67–76.

[25] Suresh, B. Acetic Acid. *Chemicals Economic Handbook.* SRI International. pp. 602. 5000. (2003).

[26] Acetic Acid: Petrochemicals: World Petrochemicals: SRI Consulting. http://chemical.ihs.com/WP/Public/Reports/acetic_acid/ (accessed Dec 18, 2011).

[27] Yoneda, N.; Kusano, S.; Yasui, M.; Pujado, P.; Wilcher, S. Recent advances in processes and catalysts for the production of acetic acid. *Applied Catalysis A, General,* 2001 221, 253-265.

[28] Acetic acid. National Institute of Standards and Technology. Retrieved 2008-02-03.

[29] Sano, K.; Uchida, H.; Wakabayashi, S. (1999). A new process for acetic acid production by direct oxidation of ethylene. *Catalysis Surveys from Asia,* 1999 3, 55-60.

[30] Otto, H.; Heinrich, E. Vinegar by submerged oxidative fermentation. *Ind. Eng. Chem.*, 1959 51, 1279-1280.

[31] Partridge, E. P. Acetic acid and cellulose acetate in the United States, a general survey of economic and technical developments. *Ind. Eng. Chem.*, 1931 23, 482-498.

[32] Hromatka, O.; Ebner, H. Investigations on vinegar fermentation: Generator for vinegar fermentation and aeration procedures. *Enzymologia,* 1949 13, 369.

[33] Sim, J. H.; Kamaruddin, A. H.; Long, W. S.; Najafpour, G. Clostridium aceticum-A potential organism in catalyzing carbon monoxide to acetic acid: Application of response surface methodology. *Enzyme and Microbial Technology,* 2007 40, 1234-1243.

[34] Fumigating Comb. Fera National Bee Unit. Retrieved 8 November 2012.

[35] Auxins page on *www.plant-hormones.info*-a website set up at Long Ashton Research Station an institute of the BBSRC (site is now on independent server).

[36] Simon, S.; Petrášek, P. Why plants need more than one type of auxin. *Plant Sci.,* 2011 180, 454-460.

[37] Taiz, L.; Zeiger, E. *Plant Physiology.* 2nd edition. Massachusetts: Sinauer Associates; 1998.

[38] Friml, J. Auxin transport-shaping the plant. *Curr. Opin. Plant Biol.,* 2003 6, 7-12.

[39] Benková, E.; Michniewicz, M.; Sauer, M.; Teichmann, T.; Seifertová, D.; Jürgens, G.; Friml, J. Local, efflux-dependent auxin gradients as a common module for plant organ formation. *Cell,* 2003 115, 591-602.

[40] Hardtke, C. S. Transcriptional auxin-brassinosteroid crosstalk: who's talking?. *BioEssays,* 2007 29, 1115-1123.

[41] Jungmook, K.; Harter, K.; Theologis, A. Protein-protein interactions among the Aux/IAAproteins. *Proc. Natl. Acad. Sci. Biochem.,* 1997 94, 11786-11791.

[42] Abel, S.; Theologis, A. Early genes and auxin action. *Plant Physiol.,* 1996 111, 9-17.

[43] Dharmasiri, N.; Dharmasiri, S.; Estelle, M. The F-box protein TIR1 is an auxin receptor. *Nature,* 2005 435, 441-445.

[44] Delker, C.; Raschke, A.; Quint, M. Auxin dynamics: the dazzling complexity of a small molecule's message. *Planta,* 2008 227, 929-941.

[45] Petrásek, J.; Mravec, J.; Bouchard, R.; Blakeslee, J. J.; Abas, M.; Seifertová, D.; Wisniewska, J.; Tadele, Z.; Kubes, M.; Covanová, M.; Dhonukshe, P.; Skupa, P.; Benková, E.; Perry, L.; Krecek, P.; Lee, O. R.; Fink, G. R.; Geisler, M.; Murphy, A. S.; Luschnig, C.; Zazímalová, E.; Friml, J. PIN proteins perform a rate-limiting function in cellular auxin efflux. *Science,* 2006 312, 914-918.

[46] Sabatini, S.; Beis, D.; Wolkenfelt, H.; Murfett, J.; Guilfoyle, T.; Malamy, J.; Benfey, P.; Leyser, O.; Bechtold, N.; Weisbeek, P.; Scheres, B. An auxin-dependent distal organizer of pattern and polarity in the *Arabidopsis* root. *Cell,* 1999 99, 463-472.

[47] Heisler, M. G.; Ohno, C.; Das, P.; Sieber, P.; Reddy, G. V.; Long, J. A.; Meyerowitz, E. M. Patterns of auxin transport and gene expression during primordium development revealed by live imaging of the *Arabidopsis* inflorescence meristem. *Curr. Biol.,* 2005 15, 1899-1911.

[48] Sorefan, K.; Girin, T.; Liljegren, S. J.; Ljung, K.; Robles, P.; Galván-Ampudia, C. S.; Offringa, R.; Friml, J.; Yanofsky, M. F.; Østergaard, L. A regulated auxin minimum is required for seed dispersal in *Arabidopsis*. *Nature,* 2009 459, 583-586.

[49] Chambers. *Science and Technology Dictionary*. ISBN 1-85296-10-3.

[50] Friml, J. (2012). That is why plants grow towards the light! VIB (the Flanders Institute for Biotechnology). http://www.vib.be/en/news/Pages/That-is-why-plants-grow-towards-the-light!.aspx

[51] Nemhauser, J. L.; Feldman, L. J.; Zambryski, P. C. Auxin and ETTIN in *Arabidopsis* gynoecium morphogenesis. *Development,* 2000 127, 3877-3888.

[52] Simon, S.; Petrášek, J. Why plants need more than one type of auxin. *Plant Sci.,* 2011 180, 454-460.

[53] Zhao, Y. Auxin biosynthesis and its role in plant development. *Ann. Rev. Plant Biol.,* 2010 61, 49-64.

[54] Mashiguchi, K.; Tanaka, K.; Sakai, T.; Sugawara, S.; Kawaide, H.; Natsume, M.; Hanada, A.; Yaeno, T.; Shirasu, K. et al. The main auxin biosynthesis pathway in *Arabidopsis*. *Proc. Natl. Acad. Sci. US,* 2011 108, 18512-18517.

[55] Won, C.; Shen, X.; Mashiguchi, K.; Zheng, Z.; Dai, X.; Cheng, Y.; Kasahara, H.; Kamiya, Y., et al. Conversion of tryptophan to indole-3-acetic acid by tryptophan aminotransferases of Arabidopsis and yuccas in Arabidopsis. *Proc. Natl. Acad. Sci. US,* 2011 108, 18518-18523.

[56] Sugawara, S.; Hishiyama, S.; Jikumaru, Y.; Hanada, A.; Nishimura, T.; Koshiba, T.; Zhao, Y.; Kamiya, Y.; Kasahara, H. Biochemical analyses of indole-3-acetaldoxime-dependent auxin biosynthesis in *Arabidopsis*. *Proc. Natl. Acad. Sci. US,* 2009 106, 5430-5435.

[57] Johnson, H. E.; Crosby, D. G. Indole-3-acetic Acid. *Org. Synth. Coll.,* 1964 5, 654.

[58] Majima, R.; Hoshino, T. Synthetische Versuche in der Indol-Gruppe, VI.: Eine neue Synthese von β-Indolyl-alkylaminen. *Berichte der Deutschen Chemischen Gesellschaft,* 1925 58, 2042-2046.

[59] Hopkins, W. G. *Introduction to Plant Physiology.* John Wiley and Sons; 1999.

[60] Dippy, J. F. J.; Hughes, S. R. C.; Laxton, J. W. Chemical constitution and the dissociation constants of monocarboxylic acids. Part XIV. Monomethyl-cyclohexanecarboxylic acids. *J. Chem. Soc.,* 1954 1954, 4102-4106.

[61] Morikawa, H.; Takahashi, M. US patent 6800482: Cultured cells of Australian laurel, Pittosporaceae and a method for culturing tissues by using said cultured cells. issued 10-05-2004.

[62] Tomlin, C. D. S. *The Pesticide Manual.* 14th Edition. UK: 2006.

[63] Lozovaya, V. A.; Markova, M. N.; Tarchevskii, I. A. *Fiziol. Rast.*, (Moscow) 1987 35, 97.

[64] Navalón, A.; Blanc, R.; Vilchez J. L. Determination of 1-naphthylacetic acid in commercial formulations and natural waters by solid-phase spectrofluorimetry. *Mikcrochim. Acta*, 1997 126, 33-38.

[65] Lu, Q. M.; Chen, L. H.; Lu, M. H.; Chen, G. N; Zhang, L. Extraction and analysis of auxins in plants using dispersive liquid-liquid microextraction followed by high-performance liquid chromatography with fluorescence detection. *J. Agric. Food Chem.*, 2010 58, 2763-2770.

[66] Guan, W.; Xu, P.; Wang, K.; Song, Y.; Zhang, H. Determination and study on dissipation of 1-naphthylacetic acid in garlic and soil using high performance liquid chromatography-tandem mass spectrometry. *Food Chem. Toxicol.*, 2011 49:2869-2874.

[67] Benotti, M. J.; Lee, F. P.; Rieger, R. A.; Iden, C. R.; Heine, C. E.; Brownawell, B. J. HPLC/TOF-MS: an alternative to LC/MS/MS for sensitive and selective determination

of polar organic contaminants in the aquatic environment. In: Ferrer, I., Thurman, E. M., editors. *Liquid Chromatography/Mass Spectrometry, MS/MS and Time of Flight MS*. New York: American Chemical Society; 2003; 109-127.

[68] ChemIndustry site

[69] 24d.org Industry Task Force II on 2,4-D Research Data

[70] Quastel, J. H. 2,4-dichlorophenoxyacetic acid (2,4-D) as a selective herbicide. *Agricultural Control Chemicals (Washington, D.C.: American Chemical Society)*, 1950 45, 244-249.

[71] Wright, T. R.; Shan, G.; Walsh, T. A.; Lira, J. M.; Cui, C.; Song, P.; Zhuang, M.; Arnold, N. L.; Lira, J. M.; Cui, C.; Song, P.; Zhuang, M.; Arnold, N. L.; Lin, G.; Yau, K.; Russell, S. M.; Cicchillo, R. M.; Peterson, M. A.; Simpson, D. M.; Zhou, N.; Ponsamuel, J.; Zhang, Z. Robust crop resistance to broadleaf and grass herbicides provided by aryloxyalkanoate dioxygenase transgenes. *Proceedings of the National Academy of Sciences,* 2010 107, 20240-20245.

[72] Pollack, A. Dow Weed Killer, Nearing Approval, Runs Into Opposition. *The New York Times*, Retrieved April 25, 2012.

[73] 2,4-Dichlorophenoxyacetic Acid (2,4-D). *International Programme on Chemical Safety*. Retrieved 22 June 2010.

[74] Dichlorophenoxyacetic acid, 2,4- (2,4-D): environmental aspects (EHC 84, 1989). *United Nations Environment Programme, the International Labour Organisation and the World Health Organization*. Archived from the original on 29 September 2010. Retrieved 24 September 2010.

[75] Hardell, L.; Walker, M. J.; Walhjalt, B.; Friedman, L. S.; Richter, E. D. Secret ties to industry and conflicting interests in cancer research. *Am. J. Ind. Med.,* 2007 50, 227-233.

[76] Leonard, C.; Burke, C. M.; O'Keane, C.; Doyle, J. S. Golf ball liver: agent orange hepatitis. *Gut,* 1997 40, 687-688.

[77] Johnston, S.; McCusker, G.; Tobinson, T. J. Golf ball liver: a cause of chronic hepatitis? *Gut,* 1998 42, 143-146.

[78] 2,4-dichlorophenol - toxicity, ecologicial toxicity and regulatory information

[79] The Monsanto Investigation

[80] Record in the Household Products Database of NLM

[81] Private Property Pesticide By-laws In Canada

[82] IARC monographs on the evaluation of carcinogenic risks to humans: An updating of IARC Monographs volumes 1 to 42. Supplement 7, WHO, Lyon, France 1987.

[83] Zahm, S. H.; Weisenburger, D. D.; Babbitt, P. A.; Saal, R. C.; Vaught, J. B.; Cantor, K. P.; Blair, A. A case-control study of non-Hodgkin's lymphoma and the herbicide 2,4-dichlorophenoxyacetic acid (2, 4-D) in Eastern Nebraska. *Epidemiology,* 1990 1, 349-356.

[84] Ibrahim, M. A.; Bond, G. G.; Burke, T. A.; Cole, P.; Dost, F. N.; Enterline, P. E.; Gough, M.; Greenberg, R. S.; Halperin, W. E.; McConnell, E.; Munro, I. C.; Swenberg, J. A.; Zahm, S. H.; Graham, J. D. Weight of the evidence on the human carcinogenicity of 2,4-D. *Environ. Health Perspect.,* 1991 96, 213-222.

[85] Burns, C. J.; Beard, K. K.; Cartmill, J. B. Mortality in chemical workers potentially exposed to 2,4-dichlorophenoxyacetic acid (2,4-D) 1945-1994: an update. *Occupational and Environmental Medicine,* 2001 58, 24-30.

[86] EPA: Federal Register: 2,4-D, 2,4-DP, and 2,4-DB; Decision Not to Initiate Special Review
[87] US EPA 2,4-D Reregistration Eligibility Decision, 2006.
[88] Lerda, D.; Rizzi, R. Study of reproductive function in persons occupationally exposed to 2,4-dichlorophenoxyacetic acid (2,4-D). *Mutat. Res.,* 1991 262, 47-50.
[89] Jones, D. C.; Miller, G. W. The effects of environmental neurotoxicants on the dopaminergic system: A possible role in drug addiction. *Biochem. Pharmacol.,* 2008 76, 569-581.
[90] Roberts, D. M.; Buckley, N. A. Urinary alkalinisation for acute chlorophenoxy herbicide poisoning. In: Darren, R. M., editor. *Cochrane Database Syst. Rev.* (1): CD00 54 88, 2007.
[91] Forest Service, Pesticide Background Statements, Vol. I Herbicides. United States Department of Agriculture, *Agriculture Handbook No. 633*, 1984.
[92] National Research Council Canada. Phenoxy Herbicides - Their Effects on Environmental Quality with Accompanying Scientific Criteria for 2,3,7,8-Tetrachlorodibenzo-p-Dioxin (TCDD). Subcommittee on Pesticides and Related Compounds, NRC Associate Committee on Scientific Criteria for Environmental Quality, Ottawa, Canada, 1978.
[93] Cavalca, L.; Hartmann, A.; Rouard, N.; Soulas, G. Diversity of tfdC genes: distribution and polymorphism among 2,4-dichlorophenoxyacetic acid degrading soil bacteria. *FEMS Microbiology Ecology,* 1999 29, 45-58.
[94] Suwa, Y.; Wright, A. D.; Fukimori, F.; Nummy, K. A.; Hausinger, R. P.; Holben, W. E.; Forney, L. J. Characterization of a chromosomally encoded 2,4-dichlorophenoxyacetic acid alpha-ketoglutafate dioxygenase from *Burkholderia sp.* strain RASC. *Applied and Environmental Microbiology,* 1996 62, 2464-2469.
[95] Cupples, A. M.; Sims, G. K. Identification of in situ 2,4-dichlorophenoxyacetic acid-degrading soil microorganisms using DNA-stable isotope probing. *Soil Biol. Biochem.,* 2007 39, 232-238.
[96] Radajewski, S.; Ineson, P.; Parekh, N. R.; Murrell, J. C. Stable-isotope probing as a tool in microbial ecology. *Nature,* 2000 403, 646-649.
[97] Howard, Philip H. *Handbook of Environmental Fate and Exposure Data for Organic Chemicals.* Lewis Publishers Chelsea, Michigan.
[98] EUROPA - Plant Health - Plant Protection - Evaluation and Authorisation - Existing active substances - Reports
[99 ttp://sv.wikipedia.org/wiki/2,4-diklorfenoxiättiksyra
[100] http://www.mddep.gouv.qc.ca/pesticides/permis-en/code-gestion-en/espace-vert.htm
[101] http://www.ene.gov.on.ca/en/news/2009/030401.php
[102] 2,4-D (2,4-dichlorophenoxyacetic acid) | Reregistration | Regulating Pesticides | Pesticides | US EPA
[103] Proposed Acceptability for Continuing Registration PACR2007-06
[104] Petition to Revoke All Tolerances and Cancel All Registrations for the Pesticide 2,4-Dichlorophenoxyacetic Acid (2,4-D); Notice of Availability (Docket Folder Summary). *Docket ID: EPA-HQ-OPP-2008-0877 Agency: EPA.* United States Environmental Protection Agency. April 7, 2012. Retrieved September 12, 2012.

[105] 2,4-D; Order Denying NRDC's Petition To Revoke Tolerances (Order). *Federal Register* 77 (75 (Wednesday, April 18, 2012)): 23135–23158. April 18, 2012. Retrieved September 12, 2012.

[106] Petitions to Revoke Tolerances; Denials: Natural Resources Defense Council, 2,4-dichlorophenoxyacetic acid (2-4D) Document ID: EPA-HQ-OPP-2008-0877-0446 Document Type: Rule Docket ID: EPA-HQ-OPP-2008-0877.

[107] Li, S. W.; Xue, L.; Xu, S.; Feng, H.; An, L. Mediators, genes and signaling in adventitious rooting. *Botanical Reviews,* 2009 75, 230-247.

[108] Overvoorde, P.; Fukaki, H.; Beeckman, T. Auxin control of root development. *Cold Spring Harb. Perspect. Biol.*, 2010 2, a001537.

[109] Debnath, S. C. Micropropagation of lingonberry: influence of genotype, explant orientation, and overcoming TDZ-induced inhibition of shoot elongation using zeatin. *HortScience,* 2005 40, 185-188.

[110] Litwinczuk, W.; Szczerba, G.; Wrona, D. Field performance of highbush blueberries (*Vaccinium* × *corymbosum* L.) cv. 'Herbert' propagated by cuttings and tissue culture. *Sci. Hort.,* 2005 106, 162-169.

[111] Meiners, J.; Schwab, M.; Szankowski, I. (2007): Efficient *in vitro* regeneration systems for *Vaccinium* species. *Plant Cell Tiss. Org. Cult.,* 2007 676, 169-176.

[112] Ludwig-Müller, J. Indole-3-butyric acid in plant growth and development. *Plant Growth Regul.,* 2000 32, 219-230.

[113] Štefančič, M.; Štampar, F.; Veberič, R.; Osterc, G. The levels of IAA, IAAsp and some phenolics in cherry rootstock 'Gisela 5' leafy cuttings pretreated with IAA and IBA. *Sci. Hort.,* 2007 112, 399-405.

[114] De Klerk, G. J.; Der Krieken, W.; De Jong, J. The formation of adventitious roots: new concepts, new possibilities. *In Vitro Cell. Dev. Biol.-Plant,* 1999 35, 189-199.

[115] Al Malki, A. A. H. S.; Elmeer, K. M. S. Effect of medium strength and charcoal combined with IBA and NAA on root initiation of *Ficus Anastasia*. *Acad. J. Plant. Sci.* 2009 2,169-172.

[116] Mihaljević, S.; Salopek-Sondi, B. Alanine conjugate of indole-3-butyric acid improves rooting of highbush blueberries. *Plant Soil Environ.*, 2012 58, 236-241.

[117] Arnold, N. P.; Binns, M. R.; Cloutier, D. C.; Barthakur, N. N.; Pellerin, R. Auxins, salt concentrations, and their interactions during *in vitro* rooting of winter-hardy and hybrid tea roses. *HortScience,* 1995 30, 1436-1440.

[118] Blakesley, D. G.; Weston, D.; Hall, J. F. The role of endogenous auxin in root initiation. Part I. Evidence from studies on auxin application and analysis of endogenous levels. *Plant Growth Regul.,* 1991 10, 341-353.

[119] De Klerk, G. J. J.; Brugge, T.; Marinova, S. Effectiveness of indole-3-acetic acid, indole-3-butyric acid, α-napthalenacetic acid during adventitious root formation *in vitro* in Malus. *Plant Cell Tiss. Org. Cult.*, 1997 49, 39-44.

[120] Samant, P. K. S.; Dash, D. K.; Mishra, H.; Dae, P. Studies on anatomical changes during rooting in stem cuttings of *Calliandra haematoxylon* and *Pentas carnea* by application of IBA. *Orissa J. Hort.,* 1990 18, 12-17.

[121] Baraldi, R.; Bertazza, G.; Predieri, S.; Bregoli, A. M.; Cohen, J. D. Uptake and metabolism of indole-3-butyric acid during the *in vitro* rooting phase in pear cultivars (*Pyrus communis*). *Acta Hort.,* 1993 329, 289-291.

[122] Smulders, M. J. M.; Van-de-Ven, E. T. W. M.; Croes, A. F.; Wullems, G. J. Metabolism of 1-naphthaleneacetic acid in explants of tobacco: Evidence for release of free hormone from conjugates. *J. Plant Growth Regul.*, 1990 9, 27-34.

[123] Kim, Y. S.; Hahn, E. J.; Yeung, E. C.; Paek, K. Y. Lateral root development and saponin accumulation as affected by IBA or NAA in adventitious root cultures of *Panax ginseng* C.A. Meyer. *In Vitro Cell. Dev. Biol.-Plant,* 2003 39, 245-249.

[124] Hossain, S. N.; Munshi, M. K.; Islam, M. R.; Hakim, L.; Hossain, M. *In vitro* propagation of plum *(Zyziphus jujube* Lam.). *Plant Cell Tiss. Org. Cult.,* 2003 13, 81-84.

[125] Sreekumar, S.; Seeni, S.; Pushpangadan, P. Micropropagation of *Hemidesmus indicus* for cultivation and production of 2-hydroxy-4-methoxybenzaldehyde. *Plant Cell Tiss. Org. Cult.,* 2000 62, 211-218.

[126] Mala, J.; Bylinsky, V. Micropropagation of endangered species *Daphne cneorum. Biol. Plant.*, 2004 48, 633-636.

[127] Kannan, P.; Premkumar, A.; Ignacimuthu, S. Organogenesis from stem explants of *Caesalpinia bonduc. J. Trop. Med. Plants,* 2006 7, 95-100.

[128] Gantait, S.; Mandal, N.; Das, P. K. Impact of auxins and activated charcoal on *in-vitro* rooting of *Dendrobium chrysotoxum* Lindl. Cv. Golden Boy. *J. of Tropical Agric.*, 2009 47, 84-86.

[129] Tereso, S.; Miguel, C. M.; Mascarenhas, M.; Roque, A.; Trindade, H.; Maroco, J.; Oliveira, M. M. Improved *in vitro* rooting of *Prunus dulcis* Mill. cultivars. *Biol. Plant.*, 2008 52, 137-144.

[130] Nandgopal, S.; Kumari, R.; Effectiveness of auxin induced *in vitro* root culture in Chicory. *J. Central Eur. Agric.*, 2007 8, 73-80.

[131] Rugini, E.; Verma, D. C. Micropropagation of difficult-to-propagate almond (*Prunus amygdalus*, Batch) cultivar. *Plant Sci. Letters*, 1983 28, 273-281.

[132] Bourgin, J. P.; Nitsch, J. B. Production of haploid Nicotian from excised stamens. *Ann. Physiol. Veg.*, 1967 9, 377-382.

[133] Rugini, E.; Jacoboni, A.; Luppino, M. Role of basal shoot darkening and exogenous putrescine treatments on *in vitro* rooting and on endogenous polyamine changes in difficult-to root wood species. *Sci Hort.*, 1993 53, 63-72.

[134] Caboni, E. Peroxidase activity and *in vitro* rooting in an almond (*Prunus dulcis* Mill) genotype. *Acta Hort.*, 1994 373, 201-205.

[135] Caboni, E.; Damiano, A. Rooting in two almond genotypes. *Plant Sci.*, 1994 96, 163-165.

[136] Murashige, T.; Skoog, F. S. A revised medium for rapid growth and bioassay with tobacco tissue culture. *Physiol. Plant.,* 1962 15, 473- 497.

[137] Caboni, E., Tonelli, M. G.; Lauri, P.; Iacovacci, P.; Kevers, C.; Damiano, C.; Gaspar, T. Biochemical aspects of almond microcuttings related to *in vitro* rooting ability. *Biol. Plant.,* 1997 39, 91-97.

[138] Ross, G. W.; Salisbury, F. B. *Plant physiology.* 2nd Edition. Belmont, California, Wadswarth Publishiny comp. Inc.; 1978.

[139] Skirvin, R. M.; Chu, M. C.; Rukan, H. Rooting studies with *Prunus spp. in vitro. Hort Science,* 1980 15, 415.

[140] Zaied, N. S. Studies on vegetative propagation of stone fruit trees. *Ph.D. Thesis, Facult. of Agrec.*, Moshtohor, Zagazig Univ., Egypt, 1997.

[141] Saeed, W. T. *In vitro* propagation of two almond (*Prunus dulcis* Mill) cv. Bull. *Fac. of Agrec. Cairo Univ.*, 1998 49, 563-574.

[142] Harda, H.; Murai, Y. Micropropagation of Prunus mume. *Plant Cell Tiss. Org. Cult.*, 1996 46, 265-267.

[143] Kassim, N. E.; Abou Rayya, S. M.; Ibrahim, I. A.; Ali, E. A. M. Effect of rooting media with or without activated charcoal, auxin type and arginine concentrations on the rooting performance of bitter almond micro-cuttings. *J. Appl. Sci. Res.*, 2012 8, 3032-3037.

[144] Rzepka-Plevnes, D.; Kulpa, D.; Gołębiowska, D.; Porwolik, D. Effects of auxins and humic acids on *in vitro* rooting of strawberry (*Fragaria* x *ananassa* Duch.). *J. Food Agric. Environ.*, 2011 9, 592-595.

[145] Soni, M.; Thakur, M.; Modgil, M. In vitro multiplication of Merton I. 793 - An apple rootstock suitable for replantation. *Indian J. Biotechnol.* 2011 10, 362-368.

[146] De Paiva Neto, V. B.; Da Mota, T. R.; Otoni, W. C. Direct organogenesis from hypocotyl derived explants of annatto (*Bixa orellana*). *Plant Cell Tiss. Org. Cult.*, 2003 75, 159-167.

[147] Cacho, M.; Moran, M.; Herrera, M. T.; Fernandez-Tarrago, J. Morphogenesis in leaf, hypocotyls and root explants of *Digitalis thapsi* L. cultured *in vitro*. *Biom. Life Sci.*, 1991 25, 117-123.

[148] Wawrosch, C.; Zeitlhofer, P.; Grauwald, B.; Kopp, B. Effects of rooting chemicals on the establishment of micropropagated *Picrorhiza kurroa* plantlents in the greenhouse. *Acta Hort.*, 2003 616, 271-274.

[149] Tiwari, V.; Singh, B. D.; Tiwari, K. N. Shoot regeneration and somatic embryogenesis from different explants of Brahmi [*Bacopa monniera* (L.) Wettst.]. *Plant Cell Rep.*, 1998 17, 538-543.

[150] Dabski, M. 2002. Badania nad ukorzenianiem mikrosadzonek wybranych gatunków roślin ozdobnych. *Rozpr. Nauk.* AR w Lublinie, vol. 262.

[151] Świstowska, A.; Kozak, D. Wpływ auksyn na ukorzenianie mikrosadzonek i adaptacje roślin *Columnea hirta* Klotzsch et Hanst. Cz. I. W kulturze *in vitro*. *Acta Sci. Pol. Hortorum Cultus*, 2004 3, 229-238.

[152] Gopi, C.; Nataraja, S. Y.; Ponmurugan, P. In vitro multiplication of *Ocimum gratissimum* L. through direct regeneration. *Afr. J. Biotechnol.*, 2006 5, 723-726.

[153] Świstowska, A.; Kozak, D. Wpływ auksyn na ukorzenianie mikrosadzonek i adaptacje roślin *Nematanthus* × *hybridus* 'Tropicana'. *Zesz. Probl. Post. Nauk Rol.*, 2005 504, 307-312.

[154] Lobna, S.; Taha, M. M.; Ibrahim, S.; Farahat, M. M. A micropropagation protocol of *Paulownia kowakamii* through *in vitro* culture technique. *Aust. J. Basic. Appl. Sci.*, 2008 2, 594-600.

[155] Pijut, P. M.; Michler, C. H. Adventitious shoot regeneration and rooting of *Prunus serotina in vitro* cultures. *HortScience*, 2006 41, 193-201.

[156] Rout, J. R.; Sahoo, S. L.; Das, R. An attempt to conserve *Withania somnifera* (L.) Dunal - A highly essential medicinal plant, through *in vitro* callus culture. *Pak. J. Bot.*, 2011 43, 1837-1842.

[157] Baig, M. M. Q.; Hafiz, I. A.; Hussain, A.; Ahmad, T.; Abbasi, N. A. An efficient protocol for *in vitro* propagation of *Rosa gruss an teplitz* and *Rosa centifolia*. *Afr. J. Biotechnol.*, 2011 10, 4564-4573.

[158] Epstein, E.; Muller, J. L. Indole-3-butyric acid in plants: Occurrence, synthesis, metabolism and transport. *Physiol. Plant.*, 1993 88, 382-389.

[159] Ali, A.; Ahmad, T.; Abbasi, N. A.; Hafiz, I. A. Q. Effect of different concentrations of auxins on *in vitro* rooting of olive cultivar "Moraiolo". *Pak. J. Bot.*, 2009 41, 1223-1231.

[160] Benelli, C.; Fabbri, A.; Grassi, S.; Lambardi, M.; Rugini, E. Histology of somatic embryogenesis in mature tissues of olive (*Olea europaea* L.). *J. Plant Biotech.*, 2001 76, 112-119.

[161] Tanimoto, E. Regulation of root growth by plant hormones: Roles for auxin and gibberellin. *Crit. Rev. Plant. Sci.*, 2005 24, 249-265.

[162] Krieken, W. M.; Breteler, H.; Visser, M. H.; Mavridou, D. The role of the conversion of IBA into IAA on root regeneration in apple: Introduction of a test system. *Plant Cell Rep.*, 1993 12, 203-206.

[163] Liu, Z. H.; Wang, W. C.; Yen, Y. S. Root formation and indole-3-acetic acid. *Bot. Bull. Acad. Sin.*, 1998 39, 113-118.

[164] Staswick, P. E.; Serban, B.; Rowe, M. Characterization and rooting ability of indole-3-acetic acid conjugates formed during rooting of mung bean cuttings. *Plant. Physiol.*, 2005 91, 1080-1084.

[165] Nanda, R. M.; Das, P.; Rout, G. R. *In vitro* clonal propagation of *Acacia mangium* and its evaluation of genetic stability through RAPD marker. *Ann. For. Sci.*, 2004 61, 381-386.

[166] Hausman, J. F. Changes in peroxidase activity, auxin level and ethylene production during root formation by poplar shoots raised *in vitro*. *Plant Growth Regul.*, 1993 13, 263-268.

[167] El-Sharabasy, S. F.; Zaid, Z. E. Effect of plant growth regulators on the development of *ex vitro* date palm (*Phoenix dactylifera* L.) Barhi cv. plantlets. *Arab. J. Biotech.*, 2010 13, 35-46.

[168] Avci, S.; Cocu, S.; Aasim, M.; Sancak, C.; Ozcan, S. Effects of treating with auxin solutions on rooting of cuttings of sainfoin (*Onobrychis viciifolia*). *Tropical Grasslands,* 2010 44, 123-127.

[169] Sharma, T.; Modgil, M.; Thakur, M. Factors affecting induction and development of *in vitro* rooting in apple rootstocks. *Indian J. Exp. Biol.*, 2007 45, 824-829.

[170] Kharrazi, M.; Nemati, H.; Tehranifar, A.; Bagheri, A.; Sharifi, A. *In Vitro* culture of carnation (*Dianthus caryophyllus* L.) focusing on the problem of vitrification. *J. Biol. Environ. Sci.*, 2011 5, 1-6.

[171] Işıkalan, C.; Akbaş, F. A.; Namlı, S.; Tilkat, E.; Başaran, D. *In vitro* micropropagation of almond *(Amygdalus communis* L. cv. Nonpareil). *Afr. J. Biotechnol.*, 2008 7, 1875-1880.

[172] Ozudogru, E. A.; Kaya, E.; Kirdok, E.; Issever-Ozturk, S. In vitro propagation from young and mature explants of thyme (*Thymus vulgaris* and *T. longicaulis*) resulting in genetically stable shoots. *In Vitro Cell. Dev. Biol.-Plant*, 2011 47, 309-320.

[173] Monteuuis, O.; Bon, M. C. Influence of auxins and darkness on *in vitro* rooting of micropropagated shoots from mature and juvenile *Acacia mangium. Plant Cell Tiss. Org. Cult.*, 2000 63, 173-177.

[174] Rathore, J. S.; Rathore, M. S.; Singh, M.; Singh, R. P.; Shekhawat, N. S. Micropropagation of mature tree of *Citrus limon*. *Indian J. Biotechnol.,* 2007 6, 239-244.

[175] Iapichino, G.; Airò, M. Micropropagation of *Metrosideros excelsa*. *In Vitro Cell. Dev. Biol.-Plant,* 2008 44, 330-337.

[176] Nobre, J.; Santos, C.; Romano, A. Micropropagation of the Mediterranean species *Viburnum tinus*. *Plant Cell Tiss. Org. Cult.,* 2000 60, 75-78.

[177] Schoene, G.; Yeager, T. Micropropagation of sweet viburnum (*Viburnum odoratissimum*). *Plant Cell Tiss. Org. Cult.* 2005 83, 271-277.

[178] Lu, M. C. Micropropagation of *Vitis thunbergii* Sieb. et Zucc., a medicinal herb, through high-frequency shoot tip culture. *Sci. Hort.,* 2005 107, 64-69.

[179] Cheng, B.; Peterson, C. M.; Mitchell, R. J. The role of sucrose, auxin and explant source on *in vitro* rooting of seedling explants of *Eucalyptus sideroxylon*. *Plant Sci.,* 1992 87, 207-214.

[180] Foppen, F. Tables for identification of carotenoid pigments. *Chromatogr. Rev.,* 1971 14, 133-298.

[181] Isler, O.; Gutmann, H.; Solms, U. *Carotenoids*. Verlag, Basel und Stuttgart, 1971.

[182] Goodwin, T. W. *Chemistry and Biochemistry of Plant Pigments*. London: Acad Press. Vol. 1-2, 1976.

[183] Goodwin, T. W. *The Biochemistry of the Carotenoids*. Vol. 1. Plants. London New York: Chapman and Hall; 1980.

[184] Young, A. J. 1991. The photoprotective role of carotenoids in higher plants. *Physiol. Plant.,* 1991 83, 702-708.

[185] Lewin, L. R. *Physiology and Biochemistry of Algae*. London: Acad Press; 1962.

[186] Stewart, W. D. P. (Ed.) *Algal Physiology and Biochemistry*. Blackwell Sci Publ.; 1974.

[187] Naeem, M.; Bhatti, I.; Ahmad, R. H.; Ashraf, M. H. Effect of some growth hormones (GA_3, IAA and kinetin) on the morphology and early or delayed initiation of bud of lentil *(Lens culinaris Medik) Pak. J. Bot.*, 2004 36, 801-809.

[188] Awan, I. U.; Baloch, M. S.; Sadozai, N. S.; Sulemani, M. Z. Stimulatory effect of GA_3 and IAA on ripening process, kernel development and quality of rice. *Pak. J. Biol. Sci.,* 1999 2, 410-412.

[189] Ritenour, M. A.; Sutter, E. G.; William, D. M.; Saltveit, M. E. IAA content and auxiliary bud development in relation to russet spotting in harvested Iceberg lettuce. *J. Am. Soc. Hort. Sci*., 1996 121, 543-547.

[190] Olaiya, C. O.; Soetan, K. O.; Ogunkolade, N. S. Evaluation of the biochemical effects of auxins on nutritional quality of tomato (*Solanum lycopersicon*), genotype JM 94/47. *Afr. J. Food Sci.,* 2010 4, 041-045.

[191] Wei, K.; Jin, X.; Chen, X.; Wu, F.; Zhou, W.; Qiu, B.; Qiu, L.; Wang, X.; Li, C.; Zhang, G. The effect of H_2O_2 and abscisic acid (ABA) interaction on β-amylase activity under osmotic stress during grain development in barley. *Plant Physiol. Biochem.,* 2009 47, 778-784.

[192] De-La Haba, P.; Roldan, J. M.; Jimenez, F. Antagonistic effect of gibberellic acid and boron on protein and carbohydrate metabolism of soybean germinating seeds. *J. Plant Nutr*., 1985 8, 1061-1073.

[193] Joshi, G.; Shukla, A.; Shukla, A. Synergistic response of auxin and ethylene on physiology of *Jatropha curcas* L. *Braz. J. Plant Physiol*., 2011 23, 67-77.

[194] Bora, K. M.; Ganesh, R.; Mathur, S. R. Paclobutrazol delayed dark-induced senescence of mung bean leaves. *Biologia,* 2007 62, 185-188.

[195] Karataş, I.; Öztürk, L.; Erşahin, Y.; Okatan, Y. Effects of auxin on photosynthetic pigments and some enzyme activities during dark-induced senescence of Tropaeolum leaves. *Pak. J. Bot.*, 2010 42, 1881-1888.

[196] Even-Chen, Z.; Atsmon, D.; Itai, C. Hormonal aspects of senescence in detached tobacco leaves. *Physiol. Plant.,* 1978 44, 377-382.

[197] Aharoni, N. Interrelationship between ethylene and growth regulators in the senescence of lettuce leaf discs. *J. Plant Growth Regul.*, 1989 8, 309-317.

[198] Misra, A. N.; Biswall, U. C. Effects of phytohormones on chlorophyll degradation during aging of chloroplast *in vivo* and *in vitro*. *Protoplasma,* 1980 105, 1615-1602.

[199] Lavelli, V.; Hidalgo, A.; Pompei, C.; Brandolini, A. Radical scavenging activity of einkorn (*Triticum monococcum* L. subsp. *monococcum*) wholemeal flour and its relationship to soluble phenolic and lipophilic antioxidant content. *J. Cereal Sci.,* 2009 49, 319-321.

[200] Sexton, R.; Roberts, J. A. Cell biology of abscission. *Ann. Rev. Plant Physiol.,* 1982 33, 133-162.

[201] Ellis, C. M.; Nagpal, P.; Young, J. C.; Hagen, G.; Guilfoyle, T. J.; Reed, J. W. Auxin response factor 1 and auxin response factor 2 regulate senescence and floral organ abscission in *Arabidopsis thaliana*. *Development,* 2005 132, 4563-4574.

[202] Tucker, M. L.; Whitelav, C. A.; Lyssenko, N. N.; Nath, L. Functional analysis of regulatory elements in the gene promoter for abscission-specific cellulose from bean and isolation, expression, and binding affinity of three TGA-type basic leucine zipper transcription factors. *Plant Physiol.,* 2002 130, 1487-1496.

[203] Williams, J. H. H.; Williams, A. L.; Pollock, C. J.; Farrar, J. F. Regulation of leaf metabolism by sucrose. *Plant Physiol.*, 1992 39, 443-446.

[204] Czerpak, R.; Dobrzyń, P.; Krotke, A.; Kicińska, E. The effect of auxins and salicylic acid on chlorophyll and carotenoid contents in *Wolffia Arrhiza* (L.) Wimm. (Lemnaceae) growing on media of various trophicities. *Polish J. Environ. Studies*, 2002 11, 231-235.

[205] Shobha, J.; Ashwani, K., Johari, S.; Kumar, A. Influence of growth regulators on biomass and biocrude yield of *Euphorbia antisyphilitica*. Department of Botany, University of Rajasthan, Jaipur 302 004, India. *J. Physiol. Res.,* 1994 7, 65-68.

[206] Taslima, K.; Hossain, F.; Ara, U. Effect of indole-3-acetic acid (IAA) on biochemical responses of Cowpea (*Vigna unguiculata* (L.) Walp) Var. Bari Fellon-1. *Bangladesh J. Sci. Ind. Res.,* 2011 46, 77-82.

[207] Durrani, F.; Subhan, M.; Mehmood, S.; Abbas, S.; Muhammad, A.; Chaudhary, F. Stimulatory effect of NAA (naphthalene acetic acid) and BAP (benzyl aminopurine) on flowers, seeds, chlorophyll and protein-content in spinach (*Spinacia oleracea* L.). *Sarhad J. Agric.,* 2010 26, 519-526.

[208] El-Shraiy, A. M.; Hegazi, A. M. Effect of acetylsalicylic acid, indole-3-butytric acid and gibberellic acid on plant growth and yield of pea (*Pisum sativum* L.). *Aust. J. Basic Appl. Sci.,* 2009 3, 3514-3523.

[209] Hsu, S. Y.; Kao, C. H. The protective effect of free radical scavengers and metal cheaters on polyethylene glycol-treated rice leaves. *Biol. Plant.*, 2003 46, 614-619.

[210] Dalvi, U. S.; Chavan, U. D.; Kachare, D. P.; Naik, R. M. Proline metabolism in sorghum and chickpea cultivars during water stress. *J. Plant physiol.*, 2007 12, 61-69.

[211] Vamil, R.; Aniat-ul-haq; Agnihotri, R. K.; Sharma, R. Effect of certain plant growth regulators on the seedling survival, biomass production and proline content of *Bambusa arundinacea*. *Sci. Res. Rep.*, 2011 1, 44-48.

[212] Göring, H.; Plescher, F. Proline accumulation induced by weak acids and IAA in coleoptiles of wheat seedlings. *Biol. Plant.*, 1986 28, 401-406.

[213] Hameed, M.; Ashraf, M. Physiological and biochemical adaptations of *Cynodon dactylon* (L.) Pers. from the Salt Range (Pakistan) to salinity stress. *Flora,* 2008 203, 683-694.

[214] Hare, P. D.; Cress, W. A.; Van Staden, J. Proline synthesis and degradation: a model system for elucidating stress-related signal transduction. *J. Exp. Bot.*, 1999 50, 413-434.

[215] Bakrim, A.; Lamhamdi, M.; Sayah, F.; Chibi, F. Effects of plant hormones and 20-hydroxyecdysone on tomato (*Lycopersicum esculentum*) seed germination and seedlings growth. *Afr. J. Biotechnol.,* 2007 6, 2792-2802.

[216] Pesci, P.; Beffagna, N. Effects of weak acids on proline accumulation in barley leaves: a comparison between abscisic acid and isobutyric acid. *Plant Cell Environ.,* 1985 8, 129-133.

[217] Woodward, A. W.; Bartel, B. Auxin: regulation, action, and interaction. *Ann. Bot.,* 2005 95, 707-735.

[218] Vanneste, S.; Friml, J. Auxin: a trigger for change in plant development. *Cell,* 2009 136, 1005-1016.

[219] Pitts, R. J.; Cernac, A.; Estelle, M. Auxin and ethylene promote root hair elongation in Arabidopsis. *Plant J.,* 1998 16, 553-560.

[220] Rahman, A.; Hosokawa, S.; Oono, Y.; Amakawa, T.; Goto, N.; Tsurumi, S. Auxin and ethylene response interactions during Arabidopsis root hair development dissected by auxin influx modulators. *Plant Physiol.*, 2002 130, 1908-1917.

[221] Ishida, T.; Kurata, T.; Okada, K.; Wada, T. A genetic regulatory network in the development of trichomes and root hairs. *Annu. Rev. Plant Biol.,* 2008 59, 365-386.

[222] Peret, B.; De Rybel, B.; Casimiro, I.; Benkova, E.; Swarup, R.; Laplaze, L.; Beeckman, T.; Bennett, M. J. Arabidopsis lateral root development: an emerging story. *Trends Plant Sci.*, 2009 14, 399-408.

[223] Möller, B.; Weijers, D. Auxin control of embryo patterning. *Cold Spring Harb. Perspect. Biol.*, 2009 1, a001545.

[224] Ljung, K.; Bhalerao, R. P.; Sandberg, G. Sites and homeostatic control of auxin biosynthesis in Arabidopsis during vegetative growth. *Plant J.,* 2001 28, 465-474.

[225] Robert, H. S.; Friml, J. Auxin and other signals on the move in plants. *Nat. Chem. Biol.*, 2009 5, 325-332.

[226] Tsurumi, S.; Wada, S. Transport of shoot and cotyledon applied indole-3-acetic acid to *Vicia faba* root. *Plant Cell Physiol.*, 1980 21, 803-816.

[227] Swarup, R.; Friml, J.; Marchant, A.; Ljung, K.; Sandberg, G.; Palme, K.; Bennett, M. Localization of the auxin permease AUX1 suggests two functionally distinct hormone transport pathways operate in the Arabidopsis root apex. *Genes Dev.,* 2001 15, 2648-2653.

[228] Geisler, M.; Murphy, A. S. The ABC of auxin transport: the role of p-glycoproteins in plant development. *FEBS Lett.*, 2006 580, 1094-1102.

[229] Petrasek, J.; Mravec, J.; Bouchard, R.; Blakeslee, J. J.; Abas, M.; Seifertova, D.; Wisniewska, J.; Tadele, Z.; Kubes, M.; Covanova, M.; Dhonukshe, P.; Skupa, P.; Benková, E.; Perry, L.; Krecek, P.; Lee, O. R.; Fink, G. R.; Geisler, M.; Murphy, A. S.; Luschnig, C.; Zazímalová, E.; Friml, J. PIN proteins perform a rate-limiting function in cellular auxin efflux. *Science,* 2006 312, 914-918.

[230] Paponov, I. A.; Teale, W. D.; Trebar, M.; Blilou, I.; Palme, K. The PIN auxin efflux facilitators: evolutionary and functional perspectives. *Trends Plant Sci.*, 2005 10, 170-177.

[231] Petrasek, J.; Friml, J. Auxin transport routes in plant development. *Development,* 2009 136, 2675 -2688.

[232] Datta, C.; Basu, P. S. Indole acetic acid production by a Rhizobium species from root nodules of leguminous shrub, Cajanus cajan. *Microbial. Res.,* 2000 155, 123-127.

[233] Ahmad, F.; Ahmad, I.; Khan, M. S. Indole acetic acid production by the indigenous isolates of Azatobacter and fluorescent Pseudomonas in the presence and absence of tryptophan. *Turk. J. Biol.,* 2005 29, 29-34.

[234] Ghosh, S.; Basu, P. S. Production of metabolism of indole acetic acid in root nodules of *Phaseolus mungo. Microbiol. Res.,* 2006 161, 362-366.

[235] Mandal, S. M.; Mandal, K. C.; Dey, S.; Pati, B. R. Optimization of cultural and nutritional conditions for indole-3-acetic acid (IAA) production by a *Rhizobium sp.* isolated from root nodules of *Vigna mungo* (L.) Hepper. *Res. J. Microbiol.,* 2007 2, 239-246.

[236] Ernstsen, A.; Sandberg, G.; Crozier, A.; Wheeler, C. T. Endogenous indoles and the biosynthesis and metabolism of indole-3-acetic acid in cultures of *Rhizobium phaseoli. Planta,* 1987 171, 422-428.

[237] Spaepen, S.; Vanderleyden, J.; Remans, R. Indole-3-acetic acid in microbial and microorganism-plant signalling. *FEMS Microbiol. Rev.,* 2007 31, 425-448.

[238] Arshad, M.; Jr Frankenberger, W. T. Plant growth regulating substances in the rhizosphere: Microbial production and function. *Adv. Agron.,* 1998 62, 45-51.

[239] Strzelczyk, E.; Pokojska-Burdziej, A. Production of auxin and gibberellin-like substances by mycorrhizal fungi, bacteria and actinomycetes isolated from soil and mycorrhizosperme of pine (*Pinus silvestris* L.). *Plant Soil.,* 1984 81, 185-194.

[240] Woodward, A. W.; Bartel, B. Auxin: regulation, action, and interaction. *Ann. Bot.*, 2005 95, 707-735.

[241] Went, F. W.; Thimann, K. V. *Phytohormones.* New York: The Macmillan Company; 1937.

[242] Kepinski, S.; Leyser, O. Plant development: auxin in loops. *Curr. Biol.,* 2005 15, 208-210.

[243] Davies, P. J. *Plant hormones: Biosynthesis, Signal Transduction, Action!* Dordrecht, The Netherlands: Kluwer Academic Publishers; 2004.

[244] Nemhauser, J. L.; Feldman, L. J.; Zambryski, P. C. Auxin and ETTIN in Arabidopsis gynoecium morphogenesis. *Development,* 2000 127, 3877-3888.

[245] Mol, R.; Filek, M.; Machackova, I.; Matthys-Rochon, E. Ethylene synthesis and auxin augmentation in pistil tissues are important for egg cell differentiation after pollination in maize. *Plant Cell Physiol.,* 2004 45, 1396-1405.

[246] Aloni, R.; Aloni, E.; Langhans, M.; Ullrich, C. I. Role of auxin in regulating Arabidopsis flower development. *Planta,* 2006 223, 315-328.

[247] Kovaleva, L.; Zakharova, E. Hormonal status of the pollen-pistil system at the progamic phase of fertilization after compatible and incompatible pollination in *Petunia hybrida* L. *Sexual Plant Reproduction,* 2003 16, 191-196.

[248] Wu, J. Z.; Qin, Y.; Zhao, J. 2008. Pollen tube growth is affected by exogenous hormones and correlated with hormone changes in styles before and after pollination in *Torenia fournieri* L. *Plant Growth Regul.,* 2008 55, 137-148.

[249] Tanaka, Y.; Sano, T.; Tamaoki, M.; Nakajima, N.; Kondo, N.; Hasezawa, S. Cytokinin and auxin inhibit abscisic acid-induced stomatal closure by enhancing ethylene production in Arabidopsis. *J. Exp. Bot.,* 2006 57, 2259-2266.

[250] Blachova, M. Changes in the level of endogenous gibberellins and auxin proceeding the formation of adventitious roots of isolated epicotyls of pea plant. *Flora,* 1969 160, 493-499.

[251] Sugiyama, M. Organogenesis *in vitro. Sci. Hort.*, 1999 116, 61-64.

[252] Wada, S.; Tanimoto, E.; Masuda, Y. Cell elongation and metabolic turnover of the cell wall as affected by auxin and cell wall degrading enzymes. *Plant Cell Physiol.,* 1998 9, 369-376.

[253] Ludwig-Müller, J. Indole-3-butyric acid in plant growth and development. *Plant Growth Regul.,* 2000 32, 219-230.

[254] Taiz, L.; Zeiger, E. *Mineral Nutrition: Plant Physiology.* 2nd edition. Sinaver Associates Inc. Pub.; 2002

[255] Davis, T. D.; Haissig, B. E.; Sankhla, N. *Adventitious root formation in cuttings. Advances in Plant Sciences Series,* vol. 2. Portland, Oregon, US: Dioscorides Press; 1989.

[256] Nordström, A. C.; Jacobs, F. A.; Eliasson, L. Effect of exogenous indole-3-acetic acid and indole-3-butyric acid on internal levels of the respective auxins and their conjugation with aspartic acid during adventitious root formation in pea cuttings. *Plant Physiol.,* 1991 96, 856-861.

[257] Nag, S.; Saha, K.; Choudhuri, M. A. Role of auxin and polyamines in adventitious root formation in relation to changes in compounds involved in rooting. *J. Plant Growth Regul.*, 2001 20, 182-194.

[258] Bellamine, J.; Penel, C.; Greppin, H.; Gaspar, T. Confirmation of the role of auxin and calcium in the late phase of adventitious root formation. *Plant Growth Regul.,* 1998 26, 191-194.

[259] De Klerk, G. J.; van der Krieken, W.; De Jong, J. C. The formation of adventitious roots: new concepts, new possibilities. *In Vitro Cell. Dev. Biol.- Plant,* 1999 35, 189-199.

[260] Riov, J. Endogenous and exogenous auxin conjugates in rooting of cuttings. *Acta Hortic.*, 1993 329, 284-288.

[261] Caboni, E.; Tonelli, M. G.; Lauri, P.; Iacovacci, P.; Kevers, C.; Damiano, C.; Gaspar, T. Biochemical aspects of almond microcuttings related to *in vitro* rooting ability. *Biol. Plant.,* 1997 39, 91-97.

[262] Van der Krieken, W. M.; Breteler, H.; Visser, M. H. M. The effect of conversion of indolebutyric acid into indoleacetic acid on root formation on microcuttings of Malus. *Plant Cell Physiol.,* 1992 33, 709-713.

[263] Güneş, T. Peroxidase and IAA-oxidase activities during rooting in cuttings of three poplar species. *Turk. J. Bot.,* 2000 24, 97-101.

[264] Norcini, J. G.; Heuser, C. W. Changes in the level of [14C] indole-3-acetic acid and [14C] indoleacetylaspartic acid during root formation in mung bean cuttings. *Plant Physiol.,* 1988 86, 1236-1239.

[265] Mamun, M. A.; Sikdar, M. B. H.; Dipak, K. P.; Rahman, M.; Rezuanul Islam, M. D. *In vitro* micropropagation of some important sugarcane varieties of Bangladesh. *Asian J. Plant Sci.,* 2004 3, 666-669.

[266] Baskaran, P. B.; Raja, R.; Jayabalan N. Development of an *in vitro* regeneration system in Sorghum (*Sorghum bicholar* L). using root tranverse thin layers (tTCLs). *Turk. J. Bot.,* 2006 30, 1-9.

[267] Davies, P. J. 2004 Regulatory factors in hormone action: Level, location and signal transduction. In: Davies, P. J., editor. *Plant Hormones*. Dordrecht: Kluwer Academic Publishers; 2004; 16-35.

[268] Law, D. M.; Davies, P. J. Comparative indole-3-acetic acid levels in the slender pea and other pea phenotypes. *Plant Physiol.*, 1990 93, 1539-1543.

[269] Cassells, A. C. The effect of 2,3,5-tri-iodobenzoic acid on caulogenesis in callus cultures of tomato and *Pelargonium. Physiol. Plant.,* 1979 46, 159-164.

[270] Fakhrai, H.; Fakhrai, F.; Evans, P. K. *In vitro* culture and plant regeneration in *Vicia faba* subsp. *equina* (var. Spring Blaze). *J. Exp. Bot.,* 1989 40, 813-817.

[271] Clare M. V.; Collin, H. A. The production of plantlets from tissue cultures of Brussels sprout (*Brassica oleracea* L. var.*gemmifera* D.C.). *Ann. Bot.*, 1974 38, 1067-1076.

[272] Forest, J. C.; Wightman, F. Amino acid metabolism in plants. III. Purification and some properties of a multispecific aminotransferase isolated from bushbean seedlings (*Phaseolus vulgaris* L.). *Can. J. Biochem.*, 1972 50, 813-829.

[273] Purves, W. K.; Brown, H. M. Indoleacetaldehyde in cucumber seedlings. *Plant Physiol.,* 1978 61, 104-106.

[274] Miyata, S.; Suzuki, Y.; Kamisaka, S.; Masuda, Y. Effects of abscisic acid on CO_2 exchange in *Lemna gibba. Physiol. Plant.,* 1981 51, 401-406.

[275] Normanly, J.; Grisafi, P.; Fink, G. R.; Bartel, B. Arabidopsis mutants resistant to the auxin effects on indole-3-acetonitrile are defective in the nitrilase encoded by the NIT1 gene. *Plant Cell,* 1997 9, 1781-1790.

[276] Normanly, J.; Slovin, J. P.; Cohen, J. D. Auxin Metabolism. In: Davies, P. J., editor. *Plant Hormones: Biosynthesis, Signal Transduction, Action!* Dordrecht: Kluwer Academic Publishers; 2004; 36-62.

[277] Slovin, J. P.; Bandurski, R. S.; Cohen, J. D. Auxin. In: Hooykaas, P. J. J., Hall, M. A., Libbenga, K. R., editors. *Biochemistry and Molecular Biology of Plant Hormones.* Amsterdam: Elsevier Science B.V.; 1999; 115-140.

[278] Klems, M.; Truksa, M.; Machackova, I.; Eder, J.; Prochazka, S. Uptake, transport and metabolism of C14-2,4-dichlorophenoxyacetic acid (C14-2,4-D) in cucumber (*Cucumis sativus* L.) explants. *Plant Growth Regul*. 1998 26, 195-202.

[279] Hopkins, W. G. *Introduction to Plant Physiology*. New York: John Wiley and sons; 1995.

[280] Went, F. W. Wuchsstoff und Wachstum. *Rec. Trav. Bot. Neerl.,* 1928 25, 1-116.

[281] Raven, J. A. Transport of indoleacetic acid in plant cells in relation to pH and electrical potential gradients, and its significance for polar IAA transport. *New Phytol.,* 1975 74, 163-174.

[282] Goldsmith, M. H. M. The polar transport of auxin. *Annu. Rev. Plant Physiol.*, 1977 28, 439-478.

[283] Morris, D. A. Transmembrane auxin carrier systems-dynamic regulators of polar auxin transport. *Plant Growth Regul.*, 2000 32, 161-172.

[284] Rayle, D. L.; Cleland, R. E. The acid growth theory of auxin-induced cell elongation is alive and well. *Plant Physiol.*, 1992 99, 1271-1274.

[285] Funada, R.; Kubo, T.; Tabuchi, M.; Sugiyama, T.; Fushitani, M. Seasonal variations in endogenous indole-3-acetic acid and abscisic acid in the cambial region of *Pinus densiflora* Sieb. et Zucc. stems in relation to earlywood/latewood transition and cessation of tracheid production. *Holzforschung,* 2001 55, 128-134.

[286] Aloni, R. *2004 The induction of vascular tissues by auxin.* In: Davies, P. J., editor. *Plant Hormones*. Dordrecht: Kluwer Academic Publishers; 2004; 471-492.

[287] Sachs, T. Integrating cellular and organismic aspects of vascular differentiation. *Plant Cell Physiol.,* 2000 41, 649-656.

[288] Liu, C. M.; Xu, Z. H.; Chua, N. H. Polar transport is essential for the establishment of bilateral symmetry during early plant embryogenesis. *Plant Cell,* 1993 5, 621-630.

[289] Van Der Krieken, W. M.; Breteler, H.; Visser, M. H. M. Uptake and metabolism of indolebutyric acid during root formation on *Malus* microcuttings. *Acta Bot. Neerl.*, 1992 41, 435-442.

[290] Cline, M. G. The role of hormones in apical dominance-new approaches to an old problem in plant development. *Physiol. Plant.*, 1994 90, 230-237.

[291] Friml, J. Auxin transport-shaping the plant. *Curr. Opin. Plant. Biol.*, 2003 6, 1-6.

[292] Hildebrandt, A. C., Wilmar, J. C., Johns, H.; Riker, A. J. Growth of edible chlorophyllous plant tissues *in vitro*. *Am. J. Bot.*, 1963 50, 248-254.

[293] Kim, Y. S.; Hahn, E. J.; Yeung, E. C.; Paek, K. Y. Lateral root development and saponin accumulation as affected by IBA or NAA in adventitious root cultures of *Panax ginseng* CA Meyer. *In Vitro Cell. Dev. Biol.-Plant,* 2003 39, 245-249.

[294] Feher, A.; Pasternak, T.; Otvos, K.; Miskolczi, P.; Dudits, D. Induction of embryogenic competence in somatic plant cells: a review. *Biologia,* 2002 57, 5-12.

[295] Lazzeri, P. A.; Hildebrand, D. F.; Sunega, J.; Williams, E. G.; Collins, G. B. Soybean somatic embryogenesis: interactions between sucrose and auxin. *Plant Cell Rep.,* 1988 7, 517-520.

[296] Dobrev, P.; Petkov, P.; Svetleva, D.; Ivanova, A.; Djilianov, D.; Petkova, S.; Atanassov, A. Analysis of some endogenous plant hormones during induction of somatic embryogenesis in common bean (*Phaseolus vulgaris* L.). *Biotechnol. Biotechnol. Equip.,* 2001 15, 17-22.

[297] Thomas, C.; Bronner, R.; Molinier, J.; Prinsen, E.; Van Onckelen, H.; Hahne, G. Immuno-cytochemical localization of indole-3-acetic acid during induction of somatic embryogenesis in cultured sunflower embryos. *Planta,* 2002 215, 577-583.

[298] Heyser, J. W.; Dykes, T. A.; Demott, K. J.; Nabors, M. W. High frequency, long-term regeneration of rice from callus culture. *Plant Sci. Lett.*, 1983 29, 175-182.

[299] Nabors, M. W.; Heyser, J. W.; Dykes, T. A.; De Mott, K. J. Long-duration, high-frequency plant regeneration from cereal tissue cultures. *Planta,* 1983 157, 385-391.

[300] Carman, J. G.; Jefferson, N. E.; Campbell, W. F. Induction of embryogenic *Triticum aestivum* L. calli. II. Quantification of organic addenda and other culture variable effects. *Plant Cell Org. Tiss. Cult.,* 1988 12, 97-110.

[301] Przetakiewicz, A.; Orczyk, W.; Nadolska-Orczyk, A. The effect of auxin on plant regeneration of wheat, barley and triticale. *Plant Cell Tiss. Org. Cult.*, 2003 73, 245-256.

[302] Minocha, S. C. pH of the medium and the growth and metabolism of cells in culture. In: Bonga and Durzan, editors. *Cell and Tissue Culture in Forestry* Vol. 1. *General Principles and Biotechnology*. Dordrecht, Boston, Lancaster: Martinus Nijhoff Publishers; 1987; 125-141.

[303] Guan, H. Y.; De Klerk, G. J. Stem segments of apple microcuttings take up auxin predominantly via the cut surface and not via epidermal surface. *Sci. Hort.,* 2000 86, 23-32.

[304] Berthon, J. Y.; Boyer, N.; Gaspar, T. Uptake, distribution and metabolism of 2,4-dichlorophenoxyacetic acid in shoots of juvenile and mature clones of *Sequoiadendron giganteum* in relation to rooting *in vitro*. *Plant Physiol. Biochem.,* 1991 29, 355-362.

[305] Klems, M.; Truksa, M.; Machackova, I.; Eder, J.; Prochazka, S. Uptake, transport and metabolism of C14-2,4-dichlorophenoxyacetic acid (C14-2,4-D) in cucumber (*Cucumis sativus* L.) explants. *Plant Growth Regul.,* 1998 26, 195-202.

[306] Sarropoulou, V. N.; Therios, I. N.; Dimassi, K. N. Melatonin promotes adventitious root regeneration in *in vitro* shoot tip explants of the commercial sweet cherry rootstocks CAB-6P (*Prunus cerasus* L.), Gisela 6 (*P. cerasus* x *P.canescens*) and MxM 60 (*P. avium* x *P. mahaleb*). *J. Pineal Res.*, 2012 52, 38-46.

In: Acetic Acids
Editor: Angelo Basile

ISBN: 978-1-62948-217-0

Chapter 3

Structural and Vibrational Properties of Some Quinoline Acetic Acid Derivatives with Potential Biological Activities

E. Romano,[1] J. L. Pergomet,[2] J. Zinczuk[2] and S. A. Brandán[1,*]

[1]Cátedra de Química General. Instituto de Química Inorgánica.
Facultad de Bioquímica, Química y Farmacia. Universidad Nacional
de Tucumán. San Miguel de Tucumán. Tucumán, Argentina
[2]Instituto de Química Rosario (CONICET-UNR), Facultad de Ciencias
Bioquímicas y Farmacéuticas, Rosario, Santa Fé, Argentina

Abstract

Quinoline compounds are widely used in medicine due to their great variety of biological and biochemical properties, especially anti-malarial and anti-microbial activities. Recently, they have been studied for their modes of function in the inhibition of tyrosine kinases, proteasome, tubulin polymerization and DNA repair. Thus, certain quinoline derivatives also show effective anticancer activity. For these reasons, the quinoline derivatives have been the object of numerous studies from different points of view. In this chapter, a complete characterization of the synthesis, structural and vibrational properties of three quinoline acetic acid derivatives with potential biological activities are studied. The knowledge of the structural and vibrational properties is important for the design and synthesis of new and better quinolinic drugs.

In this chapter, the experimental and theoretical studies of the 2-(quinolin-8-yloxy), 2-(5-chloroquinolin-8-yloxy) and 2-(5,7-dichloroquinolin-8-yloxy) acetic acids are reported. The compounds were characterized by infrared and Raman spectroscopies in the solid phase and by nuclear magnetic resonance (^{1}HNMR and ^{13}CNMR) spectroscopy in a $CDCl_3$ solution. The density functional theory (DFT) together with the 6-31G* and 6-311++G** basis sets were used to study the structural and vibrational properties. The

[*] Corresponding author: S. A. Brandán. Cátedra de Química General. Instituto de Química Inorgánica. Facultad de Bioquímica, Química y Farmacia. Universidad Nacional de Tucumán. Ayacucho 471. 4000. San Miguel de Tucumán. Tucumán. R. Argentina. Tel.: +54-381-4247752; Fax: +54-381-4248169; E- mail: sbrandan@fbqf.unt.edu.ar.

harmonic vibrational wavenumbers for the optimized geometries were calculated at the same theory levels. For a complete assignment of all the observed bands in the vibrational spectra the DFT calculations were combined with Pulay´s scaled quantum mechanical force field (SQMFF) methodology in order to fit the theoretical wavenumbers values to the experimental ones. The force constants for the three compounds were calculated and compared among them. The calculated chemical shifts of the ^{1}HNMR and ^{13}CNMR for the compounds were obtained by GIAO Method using the B3LYP/6-311++G** method. Furthermore, the characteristics of the electronic delocalization, the topological properties of the electronic charge density together with the highest occupied molecular orbital (HOMO)-lowest unoccupied molecular orbital (LUMO) energy gaps for the three compounds were studied.

Keywords: Quinoline derivatives, vibrational spectra, molecular structure, force field, DFT calculations

Introduction

The (quinolin-8-yloxy) derivatives are the object of numerous studies because they exhibit a broad range of biological properties [1-4]. Recently, the structural studies and the complete vibrational analyses of all bands observed in the vibrational spectra of 2 (quinolin-8-yloxy) [5], 2-(5-chloroquinolin-8-yloxy) [6] and 2-(5,7-dichloroquinolin-8-yloxy) acetic acids [7] were performed combining the DFT calculations with the Pulay´s scaled quantum mechanical force field (SQMFF) methodology [8, 9]. Their corresponding structures were optimized by using the hybrid B3LYP theoretical calculations together with the 6-31G* and 6-311++G** basis sets. In those studies, the normal mode calculations for the different structures of each compound were accomplished by using a generalized valence force field (GVFF) [8, 9]. In this chapter, comparisons between the structural and vibrational properties for the three 2-(quinolin-8-yloxy) acetic acids are compared and analyzed.

Methods

Synthesis

2-(quinolin-8-yloxy) [5], 2-(5-chloroquinolin-8-yloxy) acetic acids [6] and 2-(5,7-dichloroquinolin-8-yloxy) acetic acid [7] were prepared and characterized by means of the vibrational spectra in solid phase and also by NMR spectroscopy for diluted solutions in $CDCl_3$, as reported in References [5-7].

Computational Details

The initial structures of the three compounds were modeled using the *GaussView* program [10] and, later, they were optimized at the B3LYP/6-31G* and B3LYP/6-311++G** levels of theory employing the Gaussian package program [11].

For 2-(quinolin-8-yloxy) [5], 2-(5-chloroquinolin-8-yloxy) [6] and 2-(5,7-dichloroquinolin-8-yloxy) acetic acids [7], two stable conformations were considered in accordance with the potential energy curves studied at the same levels of theory. In the three molecules, an additional planar structure was considered in accordance with that reported structure for the 8-(Carboxymethoxy)quinolinium nitrate monohydrate compound [12]. Only two of the structures for the three compounds and atoms labelling can be seen in Figure 1. The planar structures are not presented here.

The nature of the different rings and bonds of all the compounds were studied by using NBO calculations [13] with the NBO 3.1 program [14], as implemented in the Gaussian 03 package [11].

Whereas, the electronic charge density topological analyses for all the structures of the three compounds were carry out by using the AIM200 program package [15] using the Bader's atoms in molecules theory [16]. The harmonic force fields were calculated at the B3LYP/6-31G* and B3LYP/6-311++G** levels of theory and, then, transformed to natural internal coordinates by using the MOLVIB program [17].

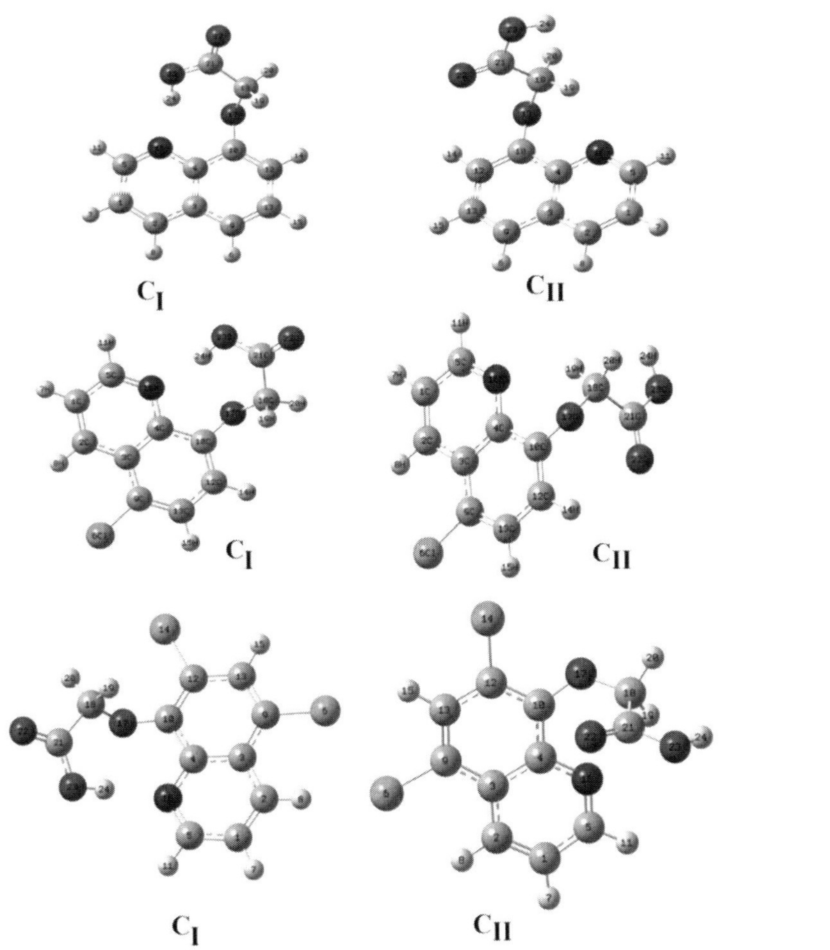

Figure 1. Theoretical structures and labelling of the atoms for the stable C_I and C_{II} conformers of 2-(quinolin-8-yloxy) (upper), 2-(5-chloroquinolin-8-yloxy) (medium) and 2-(5,7-dichloroquinolin-8-yloxy) acetic acids (bottom).

For the three compounds, the natural internal coordinates were defined approximately of the same form because the rings and the skeletal moieties are similar among them and, they are reported in the References [5-7]. To perform the complete assignments of the compounds, the harmonic force fields for the stable structures were evaluated at the B3LYP/6-31G* level employing the SQMFF procedure [8, 9] and taking into account only the potential energy distribution components (PED) ≥ 10%. For the 2-(5,7-dichloroquinolin-8-yloxy) acetic acid compound an imaginary frequency was obtained for the planar conformation and, for this reason, it structure was not considered in the vibrational analysis [7].

Results and Discussion

Geometry

The dipole moment value and the potential energy are two very important descriptors to analyze the charge distribution and the stability in a molecule. Thus, these two descriptors were calculated for the three compounds and the comparisons of these parameters are presented in Table 1. Note that the energy differences between the conformations are lower for 2-(quinolin-8-yloxy) acetic acid than the other ones while the dipole moment values are lower for the planar conformations. Obviously, the energies values increase in accordance with the addition of atoms in the compounds from 2-(quinolin-8-yloxy) to 2-(5,7-dichloroquinolin-8-yloxy) acetic acids. Moreover, the B3LYP/6-311++G** method produce the better energies values, as expected because 6-311++G** is the more complete basis set. In all the cases, the more high dipole moments are observed for the C_I conformations, as summarized in Table 1. Probably, the low dipole moment values for the planar conformation (C_{III}) of 2-(5,7-dichloroquinolin-8-yloxy) acetic acid justify the observed imaginary frequency for this compound. The stabilities and electrostatics interactions of the two conformers of each compound were investigated by using natural atomic charges [18, 19] and the results are given in Table 2. By the reasons explained above for the 2-(5,7-dichloroquinolin-8-yloxy) acetic acid only the values for the two more stable conformations are presented in the table. Notice that the natural charge values (NPA) corresponding to the C21 atoms for the 2-(5-chloroquinolin-8-yloxy) and 2-(5,7-dichloroquinolin-8-yloxy) acetic acids have higher positive values in both structures, while the more negative values correspond to the O23 atoms. Besides, the higher NPA charges observed on the O17 and O22 atoms in the C_{II} structure of 2-(quinolin-8-yloxy) acetic acid justified the repulsion between those atoms. Thus, in both forms of this compound, the C_{II} structure is more unstable than the corresponding to the C_I and C_p structures.

On the other hand, the bond orders, expressed by Wiberg's index for all conformers of the three compounds were calculated by using NBO calculations and, these values at different levels of theory are given in Table 3. For 2-(quinolin-8-yloxy)- and 2-(5-chloroquinolin-8-yloxy) acetic acids, the bond order values for the O23 atoms belonging to the COO groups in all conformers have the lowest values of the three oxygen atoms while the bond orders for the N16 atoms have higher values. The C9 and C12 atoms in both conformers of 2-(5,7-dichloroquinolin-8-yloxy) acetic acid have higher bond order values than the other ones because both atoms are linked to the Cl atoms.

Table 1. Calculated total energy (*E*) and dipolar moments for the structures of 2-(quinolin-8-yloxy), 2-(5-chloroquinolin-8-yloxy) and 2-(5,7-dichloroquinolin-8-yloxy) acetic acids at different theory levels at different theory levels

2-(quinolin-8-yloxy) acetic acid[a]							
	C_I		C_{II}			C_{III}	
Method	E_1 (Hartrees)	μ (D)	E_2 (Hartrees)	μ (D)	$\Delta E = E_1 - E_2$ (kJ/mol)	E_P (Hartrees)	μ (D)
B3LYP/6-31G*	-705.01896	7.29	-705.00384	3.99	39.66	-705.00978	2.89
B3LYP/ 6-311++G**	-705.21798	7.82	-705.20399	4.37	36.69	-705.209471	3.14
2-(5-chloroquinolin-8-yloxy) acetic acid[b]							
	C_I		C_{II}			C_{III}	
Method	E_1 (Hartrees)	μ (D)	E_2 (Hartrees)	μ (D)	$\Delta E = E_1 - E_2$ (kJ/mol)	E_P (Hartrees)	μ (D)
B3LYP/6-31G*	-1164.6146	5.56	-1164.5979	4.19	-43.80	-1164.6040	1.99
B3LYP/ 6-311++G**	-1164.8428	6.07	-1164.8251	4.38	-46.43	-1164.8305	2.08
2-(5,7-dichloroquinolin-8-yloxy) acetic acid[c]							
	C_I		C_{II}			C_{III}	
Method	E_1 (Hartrees)	μ (D)	E_2 (Hartrees)	μ (D)	$\Delta E = E_1 - E_2$ (kJ/mol)	E_P (Hartrees)	μ (D)
B3LYP/6-31G*	-1624.2030	6.55	-1624.1856	4.50	45.64	-1624.1856	0.40
B3LYP/ 6-311++G**	-1624.4561	6.79	-1624.4414	4.79	38.56	-1624.4391	0.70

[a]From Reference [5], [b]From Reference [6], [c]From Reference [7].

Table 2. Atomic charges (NPA) for all conformers of 2-(quinolin-8-yloxy), 2-(5-chloroquinolin-8-yloxy) and 2-(5,7-dichloroquinolin-8-yloxy) acetic acids at different theory levels

quinolin-8-yloxy-acetic acid[a]						
Atoms	C_I		C_{II}		C_P	
	6-31G*	6-311++G**	6-31G*	6-311++G**	6-31G*	6-311++G**
1 C	-0.275	-0.242	-0.282	-0.249	-0.275	-0.241
2 C	-0.163	-0.123	-0.168	-0.130	-0.177	-0.142
3 C	-0.084	-0.083	-0.086	-0.083	-0.084	-0.080
4 C	0.149	0.147	0.134	0.120	0.144	0.129
5 C	0.060	0.098	0.042	0.078	0.036	0.072
6 H	0.241	0.209	0.237	0.205	0.236	0.205
7 H	0.250	0.218	0.244	0.213	0.242	0.211
8 H	0.245	0.211	0.241	0.208	0.239	0.206
9 C	-0.219	-0.186	-0.224	-0.193	-0.237	-0.207
10 C	0.319	0.324	0.313	0.338	0.338	0.363
11 H	0.234	0.196	0.222	0.183	0.223	0.185
12 C	-0.269	-0.230	-0.245	-0.216	-0.328	-0.307

Table 2. (Continued)

quinolin-8-yloxy-acetic acid[a]						
Atoms	C_I		C_{II}		C_p	
	6-31G*	6-311++G**	6-31G*	6-311++G**	6-31G*	6-311++G**
13 C	-0.217	-0.181	-0.218	-0.181	-0.213	-0.178
14 H	0.251	0.220	0.261	0.229	0.239	0.210
15 H	0.246	0.212	0.243	0.209	0.240	0.206
16 N	-0.482	-0.491	-0.468	-0.473	-0.422	-0.421
17 O	-0.559	-0.578	-0.532	-0.553	-0.490	-0.505
18 C	-0.215	-0.127	-0.223	-0.144	-0.211	-0.128
19 H	0.235	0.201	0.246	0.221	0.235	0.202
20 H	0.248	0.210	0.245	0.209	0.235	0.202
21 C	0.796	0.773	0.794	0.770	0.812	0.786
22 O	-0.599	-0.593	-0.562	-0.552	-0.563	-0.555
23 O	-0.710	-0.698	-0.702	-0.679	-0.720	-0.700
24 H	0.517	0.514	0.487	0.469	0.502	0.487
2-(5-chloroquinolin-8-yloxy) acetic acid[b]						
Atoms	C_I		C_{II}		C_{III}	
	6-31G*	6-311++G**	6-31G*	6-311++G**	6-31G*	6-311++G**
1 C	-0.268	-0.234	-0.275	-0.241	-0.275	-0.233
2 C	-0.162	-0.125	-0.167	-0.131	-0.167	-0.142
3 C	-0.103	-0.125	-0.106	-0.112	-0.106	-0.108
4 C	0.159	0.160	0.145	0.134	0.145	0.142
5 C	0.064	0.101	0.046	0.082	0.046	0.076
6 Cl	0.004	0.012	-0.006	0.002	-0.006	-0.004
7 H	0.253	0.220	0.247	0.216	0.247	0.214
8 H	0.260	0.228	0.256	0.226	0.256	0.224
9 C	-0.030	0.004	-0.032	-0.024	-0.032	-0.037
10 C	0.319	0.323	0.313	0.337	0.313	0.364
11 H	0.236	0.197	0.224	0.185	0.224	0.187
12 C	-0.262	-0.220	-0.237	-0.205	-0.237	-0.296
13 C	-0.230	-0.215	-0.231	-0.203	-0.231	-0.200
14 H	0.256	0.224	0.267	0.234	0.267	0.214
15 H	0.261	0.227	0.259	0.225	0.259	0.222
16 N	-0.479	-0.488	-0.464	-0.468	-0.464	-0.418
17 O	-0.556	-0.576	-0.531	-0.552	-0.531	-0.503
18 C	-0.215	-0.127	-0.222	-0.142	-0.222	-0.129
19 H	0.236	0.202	0.245	0.219	0.245	0.202
20 H	0.250	0.212	0.246	0.210	0.246	0.202
21 C	0.796	0.772	0.794	0.769	0.794	0.786
22 O	-0.596	-0.590	-0.562	-0.552	-0.562	-0.554
23 O	-0.709	-0.696	-0.700	-0.677	-0.700	-0.698
24 H	0.517	0.513	0.488	0.469	0.488	0.488

2-(5,7-dichloroquinolin-8-yloxy) acetic acid[c]				
Atoms	C_I		C_{II}	
	6-31G*	6-311++G**	6-31G*	6-311++G**
1 C	-0.269	-0.234	-0.274	-0.241
2 C	-0.159	-0.123	-0.164	-0.128
3 C	-0.107	-0.114	-0.105	-0.111
4 C	0.169	0.163	0.153	0.145
5 C	0.068	0.105	0.050	0.085
6 Cl	0.016	0.023	0.002	0.009
7 H	0.254	0.222	0.248	0.217
8 H	0.261	0.229	0.257	0.226
9 C	-0.019	-0.010	-0.027	-0.018
10 C	0.304	0.316	0.313	0.330
11 H	0.239	0.199	0.225	0.186
12 C	-0.073	-0.070	-0.062	-0.063
13 C	-0.245	-0.226	-0.241	-0.220
14 Cl	0.035	0.042	0.033	0.042
15 H	0.275	0.241	0.272	0.238
16 N	-0.482	-0.489	-0.472	-0.475
17 O	-0.552	-0.570	-0.521	-0.540
18 C	-0.216	-0.127	-0.224	-0.143
19 H	0.237	0.203	0.238	0.200
20 H	0.254	0.217	0.244	0.208
21 C	0.796	0.772	0.797	0.772
22 O	-0.591	-0.586	-0.556	-0.545
23 O	-0.710	-0.696	-0.698	-0.673
24 H	0.515	0.512	0.489	0.470

[a]From Reference [5], [b]From Reference [6], [c]From Reference [7].

Table 3. Wiberg Indexes for all conformers of quinolin-8-yloxy-, 2-(5-chloroquinolin-8-yloxy) and 2-(5,7-dichloroquinolin-8-yloxy) acetic acids at different theory levels

quinolin-8-yloxy-acetic acid[a]						
Atoms	C_I		C_{II}		C_p	
	6-31G*	6-311++G**	6-31G*	6-311++G**	6-31G*	6-311++G**
1 C	3.941	3.959	3.943	3.960	3.944	3.961
2 C	3.939	3.958	3.942	3.959	3.944	3.961
3 C	3.992	3.996	3.993	3.996	3.993	3.996
4 C	3.983	3.984	3.987	3.991	3.992	3.996
5 C	3.915	3.932	3.928	3.945	3.929	3.946
6 H	0.943	0.960	0.946	0.961	0.946	0.962
7 H	0.939	0.955	0.942	0.957	0.943	0.958
8 H	0.942	0.958	0.944	0.960	0.945	0.961
9 C	3.944	3.961	3.945	3.961	3.945	3.961
10 C	3.918	3.914	3.918	3.913	3.909	3.903
11 H	0.947	0.964	0.953	0.969	0.952	0.969

Table 3. (Continued)

quinolin-8-yloxy-acetic acid[a]						
Atoms	C_I		C_{II}		C_p	
	6-31G*	6-311++G**	6-31G*	6-311++G**	6-31G*	6-311++G**
12 C	3.945	3.963	3.941	3.961	3.938	3.955
13 C	3.945	3.963	3.946	3.962	3.948	3.965
14 H	0.938	0.954	0.934	0.950	0.945	0.959
15 H	0.941	0.957	0.942	0.958	0.944	0.960
16 N	3.116	3.123	3.079	3.085	3.088	3.100
17 O	2.062	2.063	2.086	2.095	2.168	2.174
18 C	3.770	3.810	3.788	3.823	3.789	3.828
19 H	0.947	0.963	0.943	0.956	0.948	0.964
20 H	0.941	0.959	0.942	0.959	0.948	0.964
21 C	3.842	3.854	3.828	3.843	3.835	3.847
22 O	2.031	2.032	2.074	2.084	2.075	2.080
23 O	1.984	2.001	1.963	1.983	1.947	1.963
24 H	0.737	0.741	0.764	0.785	0.751	0.767
2-(5-chloroquinolin-8-yloxy) acetic acid[b]						
Atoms	C_I		C_{II}		C_{III}	
	6-31G*	6-311++G**	6-31G*	6-311++G**	6-31G*	6-311++G**
1 C	3.941	3.958	3.942	3.959	3.942	3.960
2 C	3.931	3.950	3.933	3.952	3.933	3.954
3 C	3.991	3.994	3.992	3.992	3.992	3.993
4 C	3.984	3.984	3.987	3.991	3.987	3.997
5 C	3.913	3.930	3.925	3.943	3.925	3.944
6 Cl	1.211	1.224	1.203	1.215	1.203	1.207
7 H	0.937	0.954	0.940	0.956	0.940	0.957
8 H	0.935	0.951	0.936	0.952	0.936	0.953
9 C	4.010	4.010	4.008	4.011	4.008	4.008
10 C	3.919	3.915	3.920	3.914	3.920	3.904
11 H	0.946	0.963	0.952	0.968	0.952	0.968
12 C	3.943	3.960	3.939	3.958	3.939	3.954
13 C	3.939	3.959	3.939	3.957	3.939	3.959
14 H	0.936	0.952	0.930	0.948	0.930	0.957
15 H	0.933	0.951	0.934	0.952	0.934	0.953
16 N	3.118	3.124	3.083	3.090	3.083	3.106
17 O	2.067	2.067	2.088	2.096	2.088	2.178
18 C	3.768	3.807	3.786	3.823	3.786	3.826
19 H	0.947	0.963	0.943	0.957	0.943	0.963
20 H	0.940	0.958	0.941	0.959	0.941	0.963
21 C	3.842	3.854	3.829	3.844	3.829	3.848
22 O	2.035	2.036	2.074	2.084	2.074	2.081
23 O	1.985	2.002	1.966	1.986	1.966	1.965
24 H	0.737	0.741	0.763	0.785	0.763	0.766

2-(5,7-dichloroquinolin-8-yloxy) acetic acid[c]				
Atoms	C_I		C_{II}	
	6-31G*	6-311++G**	6-31G*	6-311++G**
1 C	3.940	3.957	3.942	3.959
2 C	3.930	3.949	3.932	3.951
3 C	3.990	3.991	3.991	3.992
4 C	3.985	3.989	3.988	3.993
5 C	3.910	3.928	3.923	3.942
6 Cl	1.220	1.232	1.208	1.219
7 H	0.936	0.953	0.940	0.955
8 H	0.934	0.950	0.936	0.952
9 C	4.013	4.016	4.009	4.012
10 C	3.926	3.916	3.924	3.916
11 H	0.945	0.963	0.951	0.968
12 C	4.015	4.018	4.015	4.018
13 C	3.935	3.953	3.935	3.953
14 Cl	1.235	1.248	1.233	1.245
15 H	0.926	0.946	0.928	0.947
16 N	3.119	3.123	3.083	3.087
17 O	2.067	2.072	2.116	2.124
18 C	3.760	3.801	3.774	3.812
19 H	0.946	0.962	0.937	0.951
20 H	0.938	0.956	0.942	0.960
21 C	3.843	3.855	3.830	3.846
22 O	2.038	2.039	2.083	2.093
23 O	1.982	2.001	1.969	1.989
24 H	0.738	0.742	0.762	0.784

[a]From Referene [5], [b]From Reference [6], [c]From Reference [7].

The stabilities of the C_I and C_{II} structures of the three compounds were also investigated by means of NBO calculations [13, 14]. Tables 4, 5, 6 and 7 show the second order perturbation energies $E^{(2)}$ (donor → acceptor) that involve the most important delocalization for all the conformers of those compounds. For 2-(5-chloroquinolin-8-yloxy) acetic acid (Tables 5 and 6), the contributions of the stabilization energies for the $\Delta ET_{\sigma\rightarrow\sigma^*}$ charge transfers are similar to those obtained for the 2-(quinolin-8-yloxy)-acetic acid compound [5] (Table 4) while, for 2-(5,7-dichloroquinolin-8-yloxy) acetic acid, the contributions of the stabilization energies for the $\Delta ET_{\sigma\rightarrow\sigma^*}$ charge transfers are similar to those obtained for all conformers of the 2-(5-chloroquinolin-8-yloxy) acetic acid [6], while the *LP(3)Cl14* → *σ*C10-C12* delocalizations due to other Cl atom are observed for the two conformers of that dichlorinated compound (Table 7).

On the other hand, the *LP(1)N16* → *σ*O23-H24* delocalization, possible only for the C_I conformer and that confers a high stability is calculated. The calculated total energy values favour the C_I conformer, whose structure, by using both basis sets, is the most stable in the gas phase.

This way, the stabilization energies justify the different stabilities between both conformers of 2-(5,7-dichloroquinolin-8-yloxy) acetic acid.

Table 4. Main delocalization energy (in kJ/mol) for all conformers of quinolin-8-yloxy-acetic acid at different theory levels

B3LYP METHOD						
Delocalization	C_I		C_{II}		C_P	
	6-31G*	6-311++G**	6-31G*	6-311++G**	6-31G*	6-311++G**
σ C1-C2 → LP(1) C3	156.88	157.80			157.42	
σ C1-C2 → σ C5-N16*	106.51	106.17	102.70	102.37	97.02	96.18
σ C5-N16→ LP(1) C4*	167.83	167.95			178.90	
σ C5-N16 → σ C1-C2*	44.98	43.72	46.86	45.52	51.21	50.24
σ C9-C13 → LP(1) C3	175.27	175.23			168.70	
σ C9-C13→ σ C10-C12*	70.31	70.73	71.44	71.69	62.70	62.74
σ C10-C12→ LP(1) C4*	169.88	170.08			140.82	
σ C10-C12→σ C9-C13*	74.70	74.53	76.08	76.24	79.38	79.63
$\Delta ET_{\sigma\rightarrow\sigma^*}$	966.33	966.21	297.07	295.82	936.15	288.80
*LP(1)C3 → σ*C1-C2*	245.12	248.08			245.32	
*LP(1)C3 → σ*C9-C13*	220.37	222.04			219.45	
*LP(1)C4 → σ*C5-N16*	248.46	250.88			242.11	
*LP(1)C4 → σ*C10-C12*	231.32	229.77			253.02	
*LP(1)N16 → σ*O23-H24*	87.70	85.48				
*LP(2)O22 → σ*C18-C21*	90.71	82.93	89.45	84.48	92.55	86.40
*LP(2)O22 → σ*C21-O23*	129.50	121.64	147.97	143.46	146.72	145.34
*LP(2)O23 → σ*C21-O22*	130.88	195.25	184.88	169.25	192.91	176.10
$\Delta ET_{LP\rightarrow\sigma^*}$	1384.04	1436.08	422.31	397.18	1392.07	407.84
ΔETotal	2350.37	2402.29	719.38	693.00	2328.22	696.64

Table 5. Main delocalization energy (in kJ/mol) for the C_I conformer of 2-(5-chloroquinolin-8-yloxy) acetic acid at different theory levels

B3LYP METHOD[a]		
Delocalization	C_I	
	6-31G*	6-311++G**
σ C1-C2 → LP(1) C3	165.24	165.90
σ C1-C2 → σ C5-N16*	108.47	107.93
σ C5-N16→ LP(1) C4*	171.67	171.97
σ C5-N16 → σ C1-C2*	45.10	43.93
σ C9-C13 → LP(1) C3	159.26	159.80
σ C9-C13→ σ C10-C12*	63.91	64.41
σ C10-C12→ LP(1) C4*	171.21	171.59
σ C10-C12→σ C9-C13*	81.38	80.47
$\Delta ET_{\sigma\rightarrow\sigma^*}$	966.25	966.00
*LP(1)C3 → σ*C1-C2*	234.21	238.13
*LP(1)C3 → σ*C9-C13*	244.49	250.21
*LP(1)C4 → σ*C5-N16*	241.14	244.40
*LP(1)C4 → σ*C10-C12*	228.94	227.35
*LP(3)Cl6 → σ*C9 –C13*	51.75	53.50
*LP(1)N16 → σ*O23-H24*	83.68	80.59
*LP(2)O22 → σ*C18-C21*	91.17	83.31
*LP(2)O22 → σ*C21-O23*	129.91	122.18
*LP(2)O23 → σ*C21-O22*	114.53	192.70
$\Delta ET_{LP\rightarrow\sigma^*}$	1419.82	1492.39
ΔETotal	2386.07	2458.38

[a]From Reference [6].

Table 6. Main delocalization energy (in kJ/mol) for the C_{II} and C_{III} conformers of 2-(5-chloroquinolin-8-yloxy) acetic acid at different theory levels

B3LYP METHOD[a]				
Delocalization	C_{II}		C_{III}	
	6-31G*	6-311++G**	6-31G*	6-311++G**
σ C1-C2 → σ C3-C4*	63.08	63.79	63.08	66.59
σ C1-C2 → σ C5-N16*	104.75	104.17	104.75	98.27
σ C3-C4 → σ C1-C2*	74.49	75.78	74.49	73.69
σ C3-C4 → σ C5-N16*	62.99	63.87	62.99	64.29
σ C3-C4 → σ C9-C13*	69.68	71.31	69.68	69.85
σ C3-C4 → σ C10-C12*	64.62	63.70	64.62	70.35
σ C5-N16 → σ C1-C2*	46.94	45.77	46.94	50.41
σ C5-N16→ σ C3-C4*	81.47	83.01	81.47	88.16
σ C9-C13 → σ C3-C4*	68.72	69.30	68.72	67.47
σ C9-C13→ σ C10-C12*	65.00	65.33	65.00	56.97
σ C10-C12→ σ C3-C4*	73.40	74.07	73.40	58.19
σ C10-C12→σ C9-C13*	83.43	82.64	83.43	83.89
$\Delta ET_{\sigma\to\sigma^*}$	858.57	862.75	858.57	848.12
*LP(1)Cl6 → σ*C9-C13*	49.87	51.50	49.87	50.75
*LP(1)N16 → σ*C1-C5*	42.51	39.08	42.51	40.30
*LP(1)N16 → σ*C3-C4*	43.35	40.30	43.35	41.80
*LP(2)O17 → σ*C10-C12*	52.21	52.38	52.21	135.81
*LP(2)O22 → σ*C18-C21*	89.66	84.73	89.66	86.90
*LP(2)O22 → σ*C21-O23*	147.39	142.96	147.39	145.05
*LP(2)O23 → σ*C21-O22*	186.14	170.42	186.14	176.90
$\Delta ET_{LP\to\sigma^*}$	611.12	1352.77	611.12	1402.72
ΔETotal	1469.69	2215.53	1469.69	2250.85

[a]From Reference [6].

Table 7. Main delocalization energy (in kJ/mol) for the C_I conformer of 2-(5,7-dichloroquinolin-8-yloxy) acetic acid at different theory levels

B3LYP METHOD[a]		
Delocalization	C_I	
	6-31G*	6-311++G**
σ C1-C2 → LP(1) C3	167.99	168.79
σ C1-C2 → σ C5-N16*	110.56	109.73
σ C5-N16→ LP(1) C4*	174.52	175.60
σ C5-N16 → σ C1-C2*	44.43	43.51
σ C9-C13 → LP(1) C3	151.94	152.28
σ C9-C13→ σ C10-C12*	64.29	64.71
σ C10-C12→ LP(1) C4*	150.81	149.27
σ C10-C12→σ C9-C13*	76.54	75.74

Table 7. (Continued)

B3LYP METHOD[a]		
Delocalization	C_I	
	6-31G*	6-311++G**
$\Delta ET_{\sigma\to\sigma^*}$	941.09	939.62
*LP(1)C3 → σ*C1-C2*	234.16	238.85
*LP(1)C3 → σ*C9-C13*	250.13	258.28
*LP(1)C4 → σ*C5-N16*	234.75	237.93
*LP(1)C4 → σ*C10-C12*	274.42	272.37
*LP(3)Cl6 → σ*C9 –C13*	53.46	55.59
*LP(3)Cl14 → σ*C10-C12*	57.10	58.56
*LP(1)N16 → σ*O23-H24*	91.25	81.76
*LP(2)O22 → σ*C18-C21*	90.87	83.22
*LP(2)O22 → σ*C21-O23*	130.12	122.35
*LP(2)O23 → σ*C21-O22*	183.04	189.98
$\Delta ET_{LP\to\sigma^*}$	1599.31	1598.89
ΔETotal	2540.40	2538.51

[a]From Reference [7].

The electron density topological analysis for the more stable structures of the three compounds were also analysed by means of Bader's charge [16]. The calculated electron density, *(ρ)* and the *Laplacian* values, $\nabla^2\rho(r)$ in the bond critical points (BCPs) and in the ring critical points (RCPs) for those structures are shown in Tables 8, 9 and 10. The BCP has the typical properties of the closed–shell interaction and for this reason, the value of ρ(r) is relatively low, the relationship between $|\lambda_1|/\lambda_3$ and <1 $\nabla^2\rho(r)$ is positive, indicating that the interaction is dominated by the charge contraction away from the interatomic surface toward each nucleus. This analysis shows clearly for the 2-(quinolin-8-yloxy)-acetic acid compound [5] (see Table 8) two BCPs and two RCPs in the C_I structure and only one and BCP one RCP in the C_{II} structure. For the Cp structure of the compound there is not observed BCP. These results are in agreement with the NBO analysis, and demonstrate the presence of the O---H and N---H bonds in the C_I structure between the O17, N16 and H24 atoms and, moreover both BCPs in the C_I structure justify their stability. Notice that the results for the three compounds are slightly dependent on the size basis set. Thus, the analysis shows for 2-(5-chloroquinolin-8-yloxy) acetic acid two BCPs and RCPs in the C_I structure when the 6-31G* basis set is employed but, only one BCP and RCP when the 6-311++G** basis set is used. On the contrary, for the C_{II} structure with both basis sets only one BCP and RCP are observed while for the C_{III} structure there is one BCP with the 6-31G* basis set. These results, together with the NBO analysis, justify the stabilities of the C_I and C_{II} structures due to the presence of short intramolecular O17---H24 and N16---H24 bonds, in the first case and only a N16---H19 bonds in the second one, as observed in Table 9. For the dichlorinated compound, the analysis shows the same BCPs for the C_I and C_{II} structures by using both basis sets.

These results, together with the NBO analysis, justify the C_I structure stability due to the presence of a short intramolecular N16---H24 bond (1.872 Å by using 6-31G* basis set), as observed in Table 10.

Table 8. An Analysis of the Bond and Ring Critical Points (BCP, RCP) for the C_I conformer of 2-(quinolin-8-yloxy) acetic acid at different theory levels

	quinolin-8-yloxy-acetic acid[a]					
Parameter (a.u.)	6-31G*				6-311++G**	
	O17---H24	N16---H24	RCP_1	RCP_2	N16---H24	RCP_2
$\rho(r)$	0.0180	0.0357	0.0179	0.0144	0.0391	0.0139
$\nabla^2 \rho(r)$	0.0758	0.0958	0.0862	0.050	0.0970	0.0476
λ_1	-0.0194	-0.0529	-0.0180	-0.0136	-0.0613	-0.0126
λ_2	-0.0056	-0.0509	0.0072	0.0149	-0.0585	-0.0121
λ_3	0.1008	0.1997	0.0971	0.0486	0.2167	0.0481
$\|\lambda_1\| / \lambda_3$	0.1925	0.2649	0.1854	0.2798	0.2829	0.2619
Distance (Å)	2.198	1.881			1.829	

[a]From Reference [5].

Table 9. An analysis of the Bond and Ring Critical Points (BCP, RCP) for the conformers of 2-(5-chloroquinolin-8-yloxy) acetic acid at different theory levels

	B3LYP METHOD[a]											
	C_I						C_{II}				C_{III}	
Parameter	6-31G*				6-311++G**		6-31G*		6-311++G**		6-31G*	
(a.u.)	N16-H24	O17-H24	RCP_1	RCP_2	N16-H24	RCP_1	N16-H19	RCP_1	N16-H19	RCP_1	N16-H19	RCP_1
$\rho(r)$	0.0347	0.0183	0.0146	0.0182	0.0378	0.0142	0.0167	0.0119	0.0165	0.0120	0.0167	0.0119
$\nabla^2 \rho(r)$	0.0939	0.0760	0.0504	0.0889	0.0961	0.0481	0.0556	0.0581	0.0568	0.0616	0.0556	0.0582
λ_1	-0.0511	-0.0201	-0.0139	-0.0183	-0.0584	-0.0129	-0.0169	-0.0060	-0.0164	-0.0056	-0.0169	-0.0060
λ_2	-0.0491	-0.0070	0.0158	0.0093	-0.0557	0.0129	-0.0155	0.0151	-0.0148	0.0173	-0.0155	0.0151
λ_3	0.1941	0.1031	0.0485	0.0979	0.2102	0.0481	0.0880	0.0491	0.0881	0.0499	0.0880	0.0491
$\|\lambda_1\| / \lambda_3$	0.2632	0.1950	0.2860	0.1873	0.2777	0.2684	0.1925	0.1234	0.1863	0.1114	0.1925	0.1234
Distance (Å)	3.930	2.187			3.883		2.308		2.298		4.598	

[a]From Reference [6].

Also, the topological properties are more significant in this stable structure than C_{II}. This study reveals that the H bonds are different in both conformers of 2-(5,7-dichloroquinolin-8-yloxy) acetic acid. In C_I, the H bond is between N16-H24 while in C_{II} between N16-H19, as observed in Table 10. This study revels clearly the different stabilities of the conformers of each compound.

NMR Analysis

Tables 11 and 12 show a comparison between the experimental and calculated chemical shifts for the ^1H and ^{13}C nuclei of 2-(5-chloroquinolin-8-yloxy) acetic acid, respectively.

The calculated ^{1}H chemical and ^{13}C shifts show a good concordance for the three conformers when the 6-311++G** basis set is used, as expected because it is the more complete basis set [20-23].

Thus, the calculated chemical shifts for the H nuclei show a reasonable agreement in relation to experimental values with observed RMSD values between 0.578 and 0.174 ppm, while the chemical shifts for the carbon nuclei show higher RMSD values (7.325 and 0.254 ppm). Therefore, Table 12 shows that the calculated ^{13}C chemical shifts with the GIAO method using the 6-311++G** basis set are in accordance with the experimental values. In general, the calculated shifts for the ^{13}C nuclei are higher than the corresponding experimental values. Table 13 shows the ^{1}H NMR and ^{13}C NMR shifts of 2-(5,7-dichloroquinolin-8-yloxy) acetic acid in deuterated chloroform. Five signals corresponding to five different H atoms can be seen in the spectrum. Peaks at 7.64, 8.61, and 8.92 ppm belongs to three H atoms on 7, 8, and 11 positions of the hetero ring, respectively, while the peak at 7.67 belongs to H atom on position 15 of the carbocyclic ring. The peak belonging to H atoms of the CH_2 group appears at 4.77 ppm. These chemical shift values are in agreement with those reported for 2-(5-chloroquinolin-8-yloxy) acetic acid [6]. A slight shift of these peaks towards lower fields implies probably the existence of some intramolecular interaction between nonbonding electrons. In the ^{13}C NMR spectrum the peak at 170.35 ppm belongs to the C21 atom of the carboxy group and the peak at 70.85 ppm corresponds to the C18 atom of the methylene group. The observed chemical shifts are in agreement with those reported in the literature for the ^{13}C NMR of 2-(5-chloroquinolin-8-yloxy) acetic acid [6].

HOMO-LUMO Energy Gap

According to the Parr y Pearson theory [24], the frontier molecular HOMO and LUMO orbitals play an important role in the reactivity of the compounds and in many chemical properties.

Table 10. An analysis of the Bond and Ring Critical Points (BCP, RCP) for the conformers of 2-(5,7-dichloroquinolin-8-yloxy) acetic acid at different theory levels

B3LYP METHOD[a]								
Parameter (a.u.)	C_I				C_{II}			
	6-31G*		6-311++G**		6-31G*		6-311++G**	
	N16-H24	RCP_1	N16-H24	RCP_1	N16-H19	RCP_1	N16-H19	RCP_1
ρ(r)	0.0365	0.0143	0.0381	0.0142	0.0196	0.0125	0.0187	0.0125
$\nabla^2 \rho(r)$	0.0983	0.0504	0.0968	0.0481	0.0654	0.0625	0.0653	0.0648
λ_1	-0.0546	-0.0138	0.0589	0.0130	-0.0211	-0.0078	-0.0194	-0.0082
λ_2	0.0526	0.0150	-0.0563	0.0131	-0.0188	0.0170	-0.0166	0.0185
λ_3	0.2057	0.0491	0.2121	0.0481	0.1054	0.0533	0.1014	0.0545
$\|\lambda_1\| / \lambda_3$	-0.2654	-0.2811	0.2777	0.2703	-0.2002	-0.1463	-0.1913	-0.1505
Distance (Å)	1.872		1.841		2.341		2.270	

[a]From Reference [7].

The frontier molecular HOMO and LUMO orbitals were calculated for 2-(quinolin-8-yloxy), 2-(5-chloroquinolin-8-yloxy) and 2-(5,7-dichloroquinolin-8-yloxy) acetic acids and the corresponding values were compared among them, as observed in Table 14. The results show clearly that both orbitals are mainly localized on the rings, indicating that the HOMO-LUMO are mostly the π-antibonding type orbitals and that the values of the energy separation between those orbitals are higher in the 2-(quinolin-8-yloxy) acetic acid than the other ones. This large HOMO-LUMO gap for the 2-(quinolin-8-yloxy) acetic acid automatically means high excitation energies for many excited states, a good stability and a high chemical hardness. For these reasons, the presence of a Cl atom in 2-(5-chloroquinolin-8-yloxy) acetic acid and of two Cl atoms in 2-(5,7-dichloroquinolin-8-yloxy) acetic acid [7] increases their reactivities compared to the 2-(quinolin-8-yloxy) acetic acid [5].

Here, the results show clearly that the energy Gap follows the tendency: 2-(quinolin-8-yloxy) > 2-(5-chloroquinolin-8-yloxy) > 2-(5,7-dichloroquinolin-8-yloxy) acetic acids.

Table 11. Observed and calculated hydrogen chemical shifts (δ, in ppm) for 2-(5-chloroquinolin-8-yloxy) acetic acid

Atoms	C_I [a]		C_{II} [a]		C_{III} [a]		Exp [b]	
	6-31G*	6-311++G**	6-31G*	6-311++G**	6-31G*	6-311++G**	δ ppm	Multiplicity (JHz)
H19	3.65	3.98	5.63	6.45	4.15	4.46	4.92	2H s
H20	4.47	4.98	4.08	4.58	4.15	4.46		
H14	6.68	7.39	7.49	8.02	6.10	6.55	7.08	1H d (8.48)
H7	7.11	7.67	6.91	7.51	6.96	7.45	7.70	1H dd (4.18; 8.58)
H15	7.12	7.71	7.10	7.70	6.92	7.51	7.64	1H d (8.48)
H8	8.07	8.75	7.95	8.65	7.87	8.57	8.47	1H d (8.58)
H11	8.60	9.12	8.41	8.95	8.69	9.19	8.94	1H d (4.18)
H24	11.89	13.74	4.6	5.11	5.26	5.70	--	--
RMSD	0.480	0.355	0.268	0.578	0.291	0.174		

[a]GIAO Reference to TMS.
[b]Chemical shifts in $CDCl_3$ as solvent.

Table 12. Observed and calculated carbon chemical shifts (δ, in ppm) for 2-(5-chloroquinolin-8-yloxy) acetic acid

Atoms	C_I [a]		C_{II} [a]		C_{III} [a]		Exp [b]
	6-31G*	6-311++G**	6-31G*	6-311++G**	6-31G*	6-311++G**	
C18	65.8	77.5	58.8	69.3	55.3	65.1	65.94
C12	108.3	125.8	113.0	130.4	94.2	108.9	110.54
C1	109.0	126.3	107.3	124.5	108.5	125.6	123.53
C3	114.6	132.2	114.4	131.9	114.7	132.3	126.66
C13	115.4	132.4	115.1	132.1	113.6	130.2	127.07
C2	121.2	139.2	120.6	138.5	118.5	136.4	132.64
C9	122.1	140.3	121.5	139.6	118.5	136.0	121.69
C4	129.4	148.6	129.3	148.6	128.1	146.5	140.55
C5	135.0	153.4	133.4	152.1	135.0	153.8	150.28
C10	142.8	163.8	138.8	159.3	140.3	160.7	153.54
C21	150.2	172.5	150.4	171.6	148.5	170.4	170.34
RMSD	6.073	3.546	6.386	1.082	7.325	0.254	

[a]GIAO Reference to TMS.
[b]Chemical shifts in $CDCl_3$ as solvent.

Table 13. Experimental chemical shifts (in ppm) for 2-(5,7-chloroquinolin-8-yloxy) acetic acid

^{1}H NMR in $CDCl_3$[a]		
H Atoms	δ ppm	Integ. Multiplicity (JHz)
11	8.92	1H dd (4,28; 1,28)
7	7.64	1H dd (4,28; 8,73)
8	8.61	1H dd (8,73; 1,28)
15	7.67	1H s
19,20	4.77	2H s
^{13}C NMR in $CDCl_3$[a]		
C Atoms	δ ppm	Multiplicity
5	151.15	doblete
1	123.42	doblete
2	133.67	doblete
3	126.04	singlete
9	125.24	singlete
13	128.03	doblete
12	125.24	singlete
10	149.45	singlete
4	142.31	singlete
21	170.35	singlete
18	70.85	triplete

[a]From Reference [7].

Force Field

The calculated forces constants for all the structures of the three compounds studied here were calculated by using the scaling procedure of Pulay et al. [8, 9] with the MOLVIB program [17]. The results were compared among them and the values appear in Table 15. For 2-(quinolin-8-yloxy) acetic acid, the calculated $f(CH_2)$, f(C-H) and f(H-C-H) force constants for the C_I structure by using the B3LYP/6-31G* method are approximately the same for both structures while the remains force constants present variations.

Thus, the C=O and C-O stretching force constants change with the structure of the compound, being the first greater in the C_{II} structure (13.07 mdyn $Å^{-1}$) than in the C_I structure (12.53 mdyn $Å^{-1}$) because the calculated C=O distance is lower in the C_{II} structure. On the other hand, the C-O stretching force constant is greater in the C_I structure (6.73 mdyn $Å^{-1}$) because the calculated C-O distance is slightly lower in this structure. The calculated f(C-C) force constants values are in accordance with the reported values for similar molecules [18-23] as observed in Table 15.

The force constants for the two conformations of 2-(5-chloroquinolin-8-yloxy) acetic acid [6] and 2-(5,7-dichloroquinolin-8-yloxy) acetic acid [7] have approximately the same values with the exception of the f(C-Cl) force constant that is higher in the later compound, as expected because it has two Cl atoms while 2-(5-chloroquinolin-8-yloxy) acetic acid has only one [7].

Table 14. The frontier molecular HOMO and LUMO orbitals for both conformers of 2-(5,7-dichloroquinolin-8-yloxy) acetic acid

2-(5,7-dichloroquinolin-8-yloxy) acetic acid[c]				
Orbital	C_I		C_{II}	
	6-31G*	6-311++G**	6-31G*	6-311++G**
HOMO (69)	-0.24836	-0.26028	-0.22459	-0.23557
LUMO (70)	-0.09142	-0.10445	-0.07039	-0.08436
GAP (a.u.)	-0.15690	-0.15580	-0.1542	-0.15120
GAP (eV)	-4.26950	-4.23950	-4.19600	-4.11440

2-(5-chloroquinolin-8-yloxy) acetic acid[b]						
Orbital	C_I		C_{II}		C_P	
	6-31G*	6-311++G**	6-31G*	6-311++G**	6-31G*	6-311++G**
HOMO (61)	-0.24054	-0.25452	-0.22653	-0.23905	-0.22653	-0.23103
LUMO (62)	-0.08310	-0.09854	-0.06683	-0.08283	-0.06683	-0.07230
GAP (a.u.)	-0.15744	-0.15598	-0.15970	-0.15622	-0.15970	-0.15873
GAP (eV)	-4.28410	-4.24443	-4.34566	-4.25096	-4.34570	-4.31926
2-(quinolin-8-yloxy) acetic acid[a]						
Orbital	C_I		C_{II}		C_{III}	
	6-31G*	6-311++G**	6-31G*	6-311++G**	6-31G*	6-311++G**
HOMO (53)	-0.23736	-0.25313	-0.22271	-0.23632	-0.21430	-0.22804
LUMO (54)	-0.07379	-0.08994	-0.05684	-0.07351	-0.04592	-0.06262
GAP (a.u.)	-0.16357	-0.16319	-0.16587	-0.16281	-0.16838	-0.16542
GAP (eV)	-4.45090	-4.44056	-4.51349	-4.43022	-4.58179	-4.50124

[a]From Reference [5], [b]From Reference [6], [c]From Reference [7].

Table 15. Comparison of scaled internal force constants for the two conformers of the three quinoline acetic acid derivatives

B3LYP/6-31G*							
Force constant	(5,7-dichloro-quinolin-8-yloxy) acetic acid[a]		(5-chloro-quinolin-8-yloxy) acetic acid[b]			2-(quinolin-8-yloxy) acetic acid[c]	
	C_I	C_{II}	C_I	C_{II}	C_{III}	C_I	C_{II}
f(C=O)	13.65	14.22	13.62	14.16	14.10	12.53	13.07
f(C-O)	6.36	6.08	6.33	5.93	6.15	6.73	5.50
f(O-H)	6.10	7.73	6.11	7.71	7.65	5.55	7.10
f($C-H_2$)	5.34	5.29	5.31	5.23	5.10	4.87	4.80
f(C-H)	5.69	5.67	5.68	5.67	5.66	5.17	5.16
f(C-Cl)	3.65	3.61	3.51	3.49	3.42		
f(C-N)	7.89	7.76	7.90	7.77	7.84	7.26	7.12
f(C-C)	6.65	6.61	6.65	6.59	6.61	6.22	6.17
f(O=C=O)	1.34	1.23	1.33	1.24	1.26	1.33	1.23
f(H-C-H)	0.91	0.83	0.91	0.87	0.88	0.84	0.80
f(C-O-H)	1.06	0.79	1.01	0.79	0.77	0.91	0.70
f(C-O-C)	0.94	1.60	0.99	1.16	1.46	0.98	1.18

Units are mdyn $Å^{-1}$ for stretching and stretching/stretching interaction and mdyn Å rad^{-2} for angle deformations.

[a]From Reference [5], [b]From Reference [6], [c]From Reference [7].

Vibrational Analysis

The 2-(quinolin-8-yloxy), 2-(5-chloroquinolin-8-yloxy) and 2-(5,7-dichloroquinolin-8-yloxy)acetic acids structures have C_1 symmetries and 66 normal vibration modes and all the modes are active in the infrared and Raman spectra. The vibrational spectra for those compounds are presented in References [5-7]. The IR spectra of all the compounds show clearly broad bands in the 2800 and 1900 cm^{-1} region probably attributed to the N---H and O--H hydrogen bondings formed by the spatial arrangement of the molecules in the lattice crystal. In the final assignment for 2-(quinolin-8-yloxy) acetic acid only the C_I and C_P structures were considered while for 2-(5,7-dichloroquinolin-8-yloxy)acetic acid the three conformations were assigned. In accordance with the NBO and AIM results, only the C_I and C_{II} structures of 2-(5,7-dichloroquinolin-8-yloxy)acetic acid in the gas phase were considered. In the three compounds, the assignment of the experimental bands to the expected normal vibration modes were made on the basis of PED in terms of symmetry coordinates using B3LYP/6-31G* calculations, and by taking into account the corresponding assignment of related molecules [5-7, 18-26]. The complete assignments for the three compounds are summarized in References [5-7]. In this chapter, only the main bands observed in the vibrational spectra of the three compounds in the 4000-1000 cm^{-1} region can be observed in Tables 16, 17 and 18. It is important to note that the OH stretching modes in 2-(quinolin-8-yloxy) acetic acid are assigned at lower frequencies (3504 cm^{-1}) in relation to the other ones while in the dichlorinated compound those mode are observed at higher frequency (3952 cm^{-1}).

Table 16. Observed and calculated wavenumbers (cm^{-1}) and assignment for 2-(quinolin-8-yloxy)-acetic acid

Experimental[a]		C_I[a]		C_P[a]	
IR Solid	Raman Solid	SQM[b]	Assignment	SQM[b]	Assignment
	3504 w			3542	ν(O23-H24)
3105 vw	3095 w	3088	ν(C1-H7)	3094	ν(C12-H14)
3082 vw	3082 sh	3084	ν(C12-H14)	3078	ν(C1-H7)
	3070 s	3073	ν(C13-H15)	3070	ν(C13-H15)
3059 w		3059	ν(C9-H6)	3055	ν(C9-H6)
	3045 sh	3053	ν(C5-H11)	3052	ν(C2-H8)
	3025 m			3032	ν(C5-H11)
3005 vw	3013 w	3003	ν_a CH_2		
	2951 s	2943	ν_s CH_2		
2929 w				2926	ν_a CH_2
2852 vw				2892	ν_s CH_2
	1762 m	1789	ν(C21-O22)	1814	ν(C21-O22)
1639 s	1647 m	1623	ν(C3-C9)	1623	ν(C3-C9)
1604 s		1607	ν(C1-C2)		
	1598 m	1582	ν(C12-C10)		
1563 m				1576	ν(C12-C10)

Experimental[a]		C_I[a]		C_P[a]	
IR Solid	Raman Solid	SQM[b]	Assignment	SQM[b]	Assignment
1537 m	1548 w			1517	ν(C13-C12)
1455 w				1452	δCH_2
1436 w	1436 w	1433	δ CH_2		
	1364 sh	1363	ν(C3-C9)	1362	ν(C3-C4)
1342 vw	1347 w	1348	δOH	1333	ν(C5-N16)
1316 sh	1320 w	1327	ν(C5-N16)		
1308 m	1300 w			1291	δOH
1275 m	1278 m	1277	wag CH_2	1274	ν(C10-O17)
1265 sh	1266 m				
	1247 w	1251	wag CH_2, ν(C10-O17)	1254	ν(C4-N16)
1235 w	1236 m	1225	ν(C21-O23)		
1218 w		1216	ν(C2-C3), ν(C4-N16)	1218	ν(C2-C3)
1120 m	1128 m	1108	ν(C10-O17)	1111	ν(C21-O23)
1080 w	1088 m			1081	ν(C13-C9)
1050 w	1060 w	1059	ν(C13-C12)		
1044 sh	1040 vw			1031	ν(C18-O17)
1020 sh	1029 w	1029	ν(C1-C5)	1027	ν(C1-C5)
1016 w	990 w	999	ν(C18-O17)		

Abbreviations: ν, stretching; wag, wagging; δ, deformation; a, antisymmetric; s, symmetric.
[a]From Reference [5], [b]B3LYP/6-31G*.

Table 17. Observed and calculated wavenumbers (cm^{-1}) and assignment for the 2-(5-chloroquinolin-8-yloxy) acetic acid

Experimental[a]		C_I		C_{II}		C_{III}	
IR Solid	Raman Solid	SQM[b]	Assignment[a]	SQM[b]	Assignment[a]	SQM[b]	Assignment[a]
3569 w				3558	ν(O23-H24)		
3432 w.br						3541	ν(O23-H24)
3149 vvw	3145 vvw	3175	ν(O23-H24)				
3125 vvw	3130 vvw	3100	ν (C2-H8)	3109	ν (C12-H14)	3102	ν (C12-H14)
3100 w		3096	ν (C13-H15)	3097	ν (C2-H8)	3096	ν (C2-H8)
	3098 vvw	3085	ν(C1-H7)	3092	ν (C13-H15)	3086	ν (C13-H15)
3069 w	3073 w	3081	ν (C12-H14)	3079	ν(C1-H7)	3076	ν(C1-H7)
	3057 w	3056	ν (C5-H11)				
3025 vvw	3028 vw			3036	ν (C5-H11)	3037	ν (C5-H11)
	3006 vvw	3006	$ν_a$ CH_2				
	2981 vw			2974	$ν_a$ CH_2		
	2966 vw	2946	$ν_s$ CH_2				

Table 17. (Continued)

Experimental[a]		C_I		C_{II}		C_{III}	
IR Solid	Raman Solid	SQM[b]	Assignment[a]	SQM[b]	Assignment[a]	SQM[b]	Assignment[a]
	2943 vvw			2931	ν_s CH_2	2931	ν_a CH_2
2921 vw	2923 vw					2895	ν_s CH_2
1892 m,br				1823	ν (C21-O22)		
1854 m						1814	ν (C21-O22)
1696 s	1687 vw	1792	ν (C21-O22)				
1626 m	1646 vw	1613	ν (C2-C3)	1613	ν (C2-C3)	1614	ν (C2-C3)
1608 m	1610 vw	1599	ν (C1-C2)	1600	ν (C1-C2)	1601	ν (C13-C9)
1585 m	1579 m	1575	ν (C13-C9)	1570	ν (C13-C9)	1572	ν (C13-C12)
	1431 vw			1446	δCH_2	1453	δCH_2
1423 sh		1433	δCH_2			1431	wag CH_2
1395 sh	1398 m	1402	ν (C5-N16)	1403	ν (C5-N16)	1407	ν (C10-C4)
1368 s	1367 s	1353	δ OH			1350	ν (C3-C4), ν (C12-C10)
1324 s		1340	ν (C3-C4)	1336	ν (C3-C4)	1320	ν (C10-O17)
	1317 vvw			1305	ν (C12-C10)		
1280 s	1278 vvw	1305	ν (C12-C10)	1279	ρCH_2	1289	δ OH
1238 sh		1238	ν (C10-C4)	1234	ν (C10-C4)		
1042 vw	1043 w	1037	ν (C1-C5)	1036	ν (C1-C5)	1035	ν (C1-C5)
1018 m	1012 vvw			1029	ν (C18-O17)	1029	ν (C18-O17)

Abbreviations: ν, stretching; wag, wagging; δ, deformation; a, antisymmetric; s, symmetric.
[a]From Reference [6], [b]B3LYP/6-31G*.

Table 18. Observed and calculated wavenumbers (cm^{-1}) and assignment for the 2-(5,7-dichloroquinolin-8-yloxy) acetic acid

Experimental[a]		C_I[a]		C_{II}[a]	
IR Solid	Raman Solid	SQM[b]	Assignment	SQM[b]	Assignment
3952 m				3715	ν(O23-H24)
3448 w, br		3309	ν(O23-H24)		
3125		3243	ν(C13-H15)	3244	ν(C13-H15)
		3235	ν(C2-H8)	3233	ν (C2-H8)
3084 w	3081 vw	3220	ν(C1-H7)	3214	ν(C1-H7)
3041 vw	3038 vvw	3190	ν(C5-H11)	3169	ν(C5-H11)
3002 vw	3005 vvw	3152	$\nu_a CH_2$	3126	$\nu_a CH_2$
2956 w	2958 vw	3075	$\nu_s CH_2$	3069	$\nu_s CH_2$
1754 vs		1865	ν(C21-O22)	1898	ν(C21-O22)
1754 vs		1657	ν(C1-C2); ν(C2-C3)	1658	ν (C1-C2); ν(C2-C3)

Experimental[a]		C_I[a]		C_{II}[a]	
IR Solid	Raman Solid	SQM[b]	Assignment	SQM[b]	Assignment
1600 s	1601 vw	1629	ν(C13-C9)	1630	ν(C13-C9)
1580 s	1581 vw	1613	ν(C12-C10)	1608	ν(C12-C10)
1491 s		1507	δCH_2	1500	δCH_2
1455 s		1440	δOH	1437	ρCH_2
1402 s	1415 vw	1419	ν(C10-C4)	1413	ν(C10-C4)
1402 s		1392	ν(C3-C4)		
1375 s	1380 m			1380	ν(C3-C9)
1362 w	1346 vw			1332	ν (C4-N16)
1342 s		1329	ν(C4-N16)	1320	δOH
1342 s		1310	ν(C21-O23)	1310	wag CH_2
1296 w		1273	ν(C5-N16)	1277	ν(C10-O17)
1256 w		1266	wag CH_2	1260	ν(C13-C12)
1237 m		1243	ν(C13-C12)		
1144 m	1148 vw			1168	ν(C21-O23)
1101 s		1128	ν(C10-O17)	1139	ν(C18-O17)
1101 s		1076	ν(C1-C5)	1078	ν(C1-C5)
1002 m		1016	ν(C18-O17)		

Abbreviations: ν, stretching; wag, wagging; δ, deformation; ρ, rocking; a, antisymmetric; s, symmetric.
[a]From Reference [7], [b]B3LYP/6-31G*.

On the other hand, the OH deformation modes of the different conformers of the three compounds are calculated at different wavenumbers, as observed in Tables 16, 17 and 18, and for this reasons, those modes are assigned as predicted the theoretical calculations.

Another important observation is related to the antisymmetric and symmetric modes of the CH_2 groups because they are predicted by calculations in different regions for 2-(quinolin-8-yloxy) and 2-(5-chloroquinolin-8-yloxy) acetic acids while for 2-(5,7-dichloroquinolin-8-yloxy) acetic acid both modes are calculated in the same region and, for this, in this compound, for the two conformers, they are assigned to the strong IR band at 1491 cm^{-1}.

In relation to the C=O stretching modes corresponding to the carboxyl group, these modes for the different conformers of the three compounds are predicted by SQM calculations, as observed for the OH stretching, this is, at lower frequencies in 2-(quinolin-8-yloxy) acetic acid than in the other two.

Thus, in the two conformers of 2-(5,7-dichloroquinolin-8-yloxy) acetic acid those modes are assigned at 1865 and 1898 cm^{-1}; in the three conformers of 2-(5-chloroquinolin-8-yloxy) acetic acid at 1792, 1823 and 1814 cm^{-1} and, in the two conformers of 2-(quinolin-8-yloxy) acetic acid the C=O stretching modes are assigned to the bands at 1789 and 1814 cm^{-1}.

In this study, we observed that the presence of one Cl atom in 2-(5-chloroquinolin-8-yloxy) acet acid and of two Cl atoms in 2-(5,7-dichloroquinolin-8-yloxy) acetic acid justifies the shifting in some bands observed in the corresponding vibrational spectra, in reference to the 2-(quinolin-8-yloxy) acetic acid.

Conclusion

In this chapter, a comparison between the vibrational and NMR spectra of the 2-(quinolin-8-yloxy), 2-(5-chloroquinolin-8-yloxy) and 2-(5,7-dichloroquinolin-8-yloxy)acetic acids compounds were presented. The presence of Cl atoms in 2-(5-chloroquinolin-8-yloxy) and 2-(5,7-dichloroquinolin-8-yloxy) acetic acids justifies the shifting in some bands observed in the corresponding vibrational spectra. NMR spectra observed for the three compounds were successfully compared with the calculated chemical shifts at the B3LYP/6-311++G** level of theory.

On the other hand, a comparison between the three theoretical 2-(quinolin-8-yloxy), 2-(5-chloroquinolin-8-yloxy) and 2-(5,7-dichloroquinolin-8-yloxy)acetic acids structures by using B3LYP/6-31G* and B3LYP/6-311++G** calculations were obtainable. The calculations show only two stable structures for 2-(quinolin-8-yloxy) and 2-(5,7-dichloroquinolin-8-yloxy)acetic acids while for 2-(5-chloroquinolin-8-yloxy) acetic acid three stable conformers were observed. In all cases, the NBO and AIM analyses using both basis sets confirm the presence of the more stable conformers of each compound. The SQM/B3LYP/6-31G* force fields were obtained for those stable structures of 2-(quinolin-8-yloxy), 2-(5-chloroquinolin-8-yloxy) and 2-(5,7-dichloroquinolin-8-yloxy)acetic acids and also the force constants for the stretching and deformation modes.

Finally, the study of the frontier orbitals reveals clearly that the introduction of one and two Cl atoms in the quinoline ring increases the reactivities of 2-(5-chloroquinolin-8-yloxy) acetic acid in relation to 2-(quinolin-8-yloxy) acetic acid and, of 2-(5,7-dichloroquinolin-8-yloxy) acetic acid as compared to 2-(quinolin-8-yloxy) acetic acid.

Acknowledgments

This work was supported with grants from CIUNT (Consejo de Investigaciones, Universidad Nacional de Tucumán) and CONICET (Consejo Nacional de Investigaciones Científicas y Técnicas, R. Argentina). The authors thank Prof. Tom Sundius for his permission to use MOLVIB.

References

[1] S. Y. Cho, J. H. Ahn, J. D. Ha, S. K. Kang, J. Y. Baek, S. S. Han, E. Y. Shin, S. S. Kim, K. R. Kim, H. G. Cheon and J. K. Choi, "Protein Tyrosine Phosphatase 1B Inhibitors: Heterocyclic Carboxylic Acids," *Bulletin of the Korean Chemical Society*, Vol. 24, No. 10, 2003, pp. 1455-1464.

[2] B. P. Kennedy and C. Ramachandran, "Protein Tyrosine Phosphatase-1B in Diabetes," *Biochemical Pharmacology*, Vol. 60, No. 7, 2000, pp. 877-883.

[3] A. Hubele, *"Use of Quinoline Derivatives for the Protection of Cultivated Plants from Herbicides,"* US Patent No. 4902340, 1990.

[4] F. Hayat, A. Salahuddin, J. Zargan, A. Azam, Synthesis, characterization, antiamoebic activity and cytotoxicity of novel 2-(quinolin-8-yloxy) acetohydrazones and their

cyclized products (1,2,3-thiadiazole and 1,2,3-selenadiazole derivatives) *Eur. J. Med. Chem.* 45 (2010) 6127-6134.

[5] G. R. Argañaraz, E. Romano, J. Zinczuk, S. A. Brandán, Structural and vibrational study of 2-(8-quinolinyloxy)-acetic acid based on FTIR-Raman spectroscopy and DFT calculations. *J. Chem. and Chem. Eng.* 5(8) (2011) 747-758.

[6] E. Romano, M. V. Castillo, J. L. Pergomet, J. Zinczuk, S. A. Brandán, Synthesis, structural study and vibrational spectra of 2-(5-chloroquinolin-8-yloxy) acetic acid, *Open J. Synthesis, Theory and Applications*, 2 (2013) 8-22.

[7] E. Romano, M. V. Castillo, J. L. Pergomet, J. Zinczuk, S. A. Brandán, Synthesis, structural and vibrational analysis of 2-(5,7-dichloroquinolin-8-yloxy) acetic acid, *J. Mol. Struct.* 1018 (2012) 149–155.

[8] P. Pulay, G. Fogarasi, F. Pang, E. Boggs, Combination of theoretical ab initio and experimental information to obtain reliable harmonic force constants. Scaled Quantum Mechanical (SQM) force fields for glyoxal, acrolein, butadiene, formaldehyde, and ethylene. *J. Am. Chem. Soc.* 101(10) (1979) 2550.

[9] G. Rauhut and P. Pulay, "Transferable Scaling Factors for Density Functional Derived Vibrational Force Fields," *The Journal of Physical Chemistry*, Vol. 99, No. 39, 1995, pp. 3093-3100.

[10] A. B. Nielsen, A. J. Holder, *Gauss View, User's Reference*, GAUSSIAN, Inc.: Pittsburgh, PA, US 1997-1998.

[11] M. J. Frisch, C. W. Trucks, H. B. Schlegel, G. E. Scuseria, M. A. Robb, J. R. Cheeseman, J. A. Montgomery, Jr., T. Vreven, K. N. Kudin, J. C. Burant, J. M. Millam, S. S. Iyengar, J. Tomasi, V. Barone, B. Mennucci, M. Cossi, G. Scalmani, N. Rega, G. A. Petersson, H. Nakatsuji, M. Hada, M. Ehara, K. Toyota, R. Fukuda, J. Hasegawa, M. Ishida, T. Nakajima, Y. Honda, O. Kitao, H. Nakai, M. Klene, X. Li, J. E. Knox, H. P. Hratchian, J. B. Cross, C. Adamo, J. Jaramillo, R. Gomperts, R. E. Stratmann, O. Yazyev, A. J. Austin, R. Cammi, C. Pomelli, J. W. Ochterski, P. Y. Ayala, K. Morokuma, G. A. Voth, P. Salvador, J. J. Dannenberg, V. G. Zakrzewski, S. Dapprich, A. D. Daniels, M. C. Strain, O. Farkas, D. K. Malick, A. D. Rabuck, K. Raghavachari, J. B. Foresman, J. V. Ortiz, Q. Cui, A. G. Baboul, S. Clifford, J. Cioslowski, B. B. Stefanov, G. Liu, A. Liashenko, P. Piskorz, I. Komaromi, R. L. Martin, D. J. Fox, T. Keith, M. A. Al-Laham, C. Y. Peng, A. Nanayakkara, M. Challacombe, P. M. W. Gill, B. Johnson, W. Chen, M. W. Wong, C. Gonzalez, and J. A. Pople, *Gaussian 03, Revision B.01*, Gaussian, Inc., Pittsburgh PA, 2003.

[12] F. Sun, L. Chen, H. C. Fang, X. M. Lin and Y. P. Cai, "8-(Carboxymethoxy) Quinolinium Nitrate Monohydrate," *Acta Crystallographica*, Vol. 64, 2008, p. 1641.

[13] A. E. Reed, L. A. Curtis, F. Weinhold, Intermolecular interactions from a natural bond orbital, donor-acceptor viewpoint. *Chem. Rev.* 88 (6) (1988) 899-926.

[14] E. D. Glendening, J. K. Badenhoop, A. D. Reed, J. E. Carpenter, F. Weinhold, NBO 3.1; Theoretical Chemistry Institute, University of Wisconsin; Madison, WI, 1996.

[15] F. Biegler-Köning, J. Schönbohm, D. Bayles, AIM2000; A Program to Analyze and Visualize Atoms in Molecules, *J. Comput. Chem.* 22 (2001) 545.

[16] R. F. W. Bader, *Atoms in Molecules, A Quantum Theory*, Oxford University Press, Oxford, ISBN: 0198558651, 1990.

[17] T. Sundius, Scaling of Ab-initio force fields by MOLVIB. *Vib. Spectrosc.* 29 (2002) 89-95.

[18] A. E. Ledesma, J. Zinczuk, A. Ben Altabef, J. J. López-González, S. A. Brandán, Synthesis and vibrational analysis of N-(2'-furyl)-imidazole, *J. Raman Spect.* 40(8) (2009) 1004-1010.

[19] J. Zinczuk, A. E. Ledesma, S. A. Brandán, O. E. Piro, J. J. López-González, A. Ben Altabef, Structural and vibrational study of -2-(2'- furyl)-4, 5-1*H*-dihydroimidazole. *J. Phys. Org. Chem.* 22 (2009) 1166–1177.

[20] A. E. Ledesma, S. A. Brandán, J. Zinczuk, O. Piro, J. J. López González, A. Ben Altabef, Structural and vibrational study of 2-(2'- furyl)-1*H*-imidazole, *J. Phys. Chem. Org.* 21(12) (2008) 1086-1097.

[21] A. E. Ledesma, J. Zinczuk, J. J. López González, A. Ben Altabef, S. A. Brandán, Structural and vibrational study of 4-(2'- furyl)-1- methylimidazole, *J. Mol. Struc.* 924-926 (2009). 322-326.

[22] A. E. Ledesma, J. Zinczuk, J. J. López González, A. Ben Altabef, S. A. Brandán, Structural, vibrational spectra and normal coordinate analysis for two tautomers of 4(5)-(2'-furil)-1*H*-imidazol, *J. Raman Spectrosc.* 41(5) (2010) 587-597.

[23] A. E. Ledesma, C. Contreras, J. Svoboda, A. Vektariane, S. A. Brandán, Theoretical structures and experimental vibrational spectra of isomeric benzofused thieno [3,2-*b*] furan compounds, *J. Mol. Struct.*, 967 (2010) 159-165.

[24] R. G. Parr, R. G. Pearson, Absolute hardness: companion parameter to absolute electronegativity. *J. Am. Chem. Soc.* 105 (1983) 7512-7516.

[25] C. D. Contreras, A. E. Ledesma, H. E. Lanús, J. Zinczuck, S. A. Brandán, Hydration of L-tyrosine in aqueous medium. An experimental and theoretical study by mixed quantum mechanical/molecular mechanics methods, *Vibrat. Spectr.* 57 (2011) 108-115.

[26] S. A. Brandán, F. Márquez López, M. Montejo, J. J. López González, A. Ben Altabef, Theoretical and Experimental Vibrational Spectrum Study of 4-hydroxybenzoic acid as Monomer and Dimer, *Spectrochim. Acta* Part A 75 (2010) 1422-1434.

In: Acetic Acids
Editor: Angelo Basile

ISBN: 978-1-62948-217-0

Chapter 4

Use of Acetic Acid as an Acetylating Agent of Alcohols and Phenols Catalyzed by Recyclable Solid Acids

Luis R. Pizzio,* Gustavo P. Romanelli† and Mirta N. Blanco‡
Centro de Investigación y Desarrollo en Ciencias Aplicadas "Dr. Jorge J. Ronco" (CINDECA), Departamento de Química, Facultad de Ciencias Exactas, Universidad Nacional de La Plata-CCT La Plata, CONICET, 47 Nº 257, 1900 La Plata Argentina

Abstract

Acetic acid is one of the simplest carboxylic acid, and an important chemical and industrial chemical for different uses. It is employed, for example, in the production of cellulose and, in the food, diluted acetic acid acts as an acidity and as a condiment. The global demand of acetic acid is very important, around 6.5 million tonnes per year.

The acetylation reaction of organic compounds is one of the most frequently used transformations in organic synthesis. Among the various protected groups used for the hydroxyl function, the acetyl group is the most convenient in view of its easy introduction, being stable to the acidic reaction condition and also easily removable by mild alkaline hydrolysis. The most common and inexpensive reagent used to perform the acetylation is the acetic anhydride however, from an environmental point of view, a much convenient method for performing the acetylation is the use of acetic acid as acetylating agent, because only water is produced as by product, and it is a relative cheaper route for industrial purposes when compared to anhydrides or acid chlorides.

In this chapter we reviewed the performance of different recyclable solid acids synthesized in our laboratory as catalysts for the use of acetic acid as acetylating agent: 1) alcohols and phenols acetylation using borated zirconia; 2) alcohols and phenols

* Luis R. Pizzio: Centro de Investigación y Desarrollo en Ciencias Aplicadas "Dr. Jorge J. Ronco" (CINDECA), Departamento de Química, Facultad de Ciencias Exactas, Universidad Nacional de La Plata-CCT La Plata, CONICET, 47 Nº 257, 1900 La Plata, Argentina. E-mail: lrpizzio@quimica.unlp.edu.ar.

† Gustavo P. Romanelli. E-mail: gpr@quimica.unlp.edu.ar.

‡ Mirta N. Blanco. E-mail: mnblanco@quimica.unlp.edu.ar.

acetylation using borated zirconia modified with ammonium metatungstate; 3) isoamyl alcohol acetylation using tungstophosphoric acid supported on zirconia; and 4) isoamyl alcohol acetylation using partially substituted Cs and K salts of tungstophosphoric acid.

The alcohols and phenols acetylation with the borated zirconia-based catalysts gave very good yields to the desired products with very high selectivity, though the alcohol and phenol structure is a key in the yield to the acetylation product. The zirconia-supported tungstophosphoric acid prepared by the incipient wetness impregnation led to obtain isoamyl acetate with high activity and selectivity, and similar results were obtained using the Cs and K tungstophosphates, observing a correlation with the acidic properties of the catalysts.

Keywords: Acetic acid, acetylation, recyclable solid acids

List of Symbols/Acronyms

AMT	ammonium metatungstate
Ca	adsorbed concentration
C_T	total concentration
DMAP	4-(dimethylamino)pyridine
DRS	diffuse reflectance spectroscopy
DTA	differential thermal analysis
E	equilibrium adsorption impregnation
Ei	initial electrode potential
FT-IR	Fourier transform-infrared spectroscopy
GCMS	gas chromatography mass spectrometry
HPA	heteropolyacids
IEP	isoelectric point
IPA	isopropanol
MPS	mean pore size
^{31}P MAS-NMR	^{31}P magic angle spinning-nuclear magnetic resonance spectroscopy
S_{BET}	specific surface area
S_{micro}	microporous area
TG	thermogravimetric analysis
TLC	thin layer chromatography
TOF	turn over frequency
TON	turnover number
TPA	tungstophosphoric acid
V_p	pore volume
W	incipient wetness impregnation
X	number identifying the support
XRD	X-ray diffraction
Y	boron concentration
Z	zirconia

1. Introduction

Acetic acid has a very important global demand. Among the most common acetic acid uses it can be mentioned the manufacture of fragrances and perfumes, the use as a solvent for many industrial processes, the production of synthetic fibers and textiles, the manufacture of inks and dyes in clinical laboratory analysis where it is used to test blood, the fabrication of soft drink bottles, the use as a treatment for outer ear infections where it helps to prevent the growth of fungus and bacteria, the manufacture of rubbers and plastics, in the production of pesticides, and it is also approved as a food additive and has been given the E Number E260 [1].

The research about processes for functional group protection is one of the most important fields in synthetic organic chemistry. In many preparations of fragile organic compounds, some specific parts of their molecules cannot survive the required reagents or chemical environments. Then, these parts, or groups, must be protected. Protecting groups are more commonly used in small-scale laboratory work and initial development than in industrial production processes because their use adds additional steps and material costs to the process. However, the availability of a cheap building block can overcome these additional costs. The protection of functional groups is often necessary in several transformations along a synthetic sequence, particularly in the construction of polyfunctional molecules such as carbohydrates, steroids, nucleosides, and natural products [2].

The acylation of alcohols and phenols is one of the most frequently used transformations in organic synthesis as it provides an efficient and inexpensive means for protecting hydroxyl groups in a multistep synthetic process [3]. Among the various protecting groups used for the hydroxyl function, the acetyl group is the most convenient in view of its easy introduction, being stable to the acidic reaction conditions and also easily removable by mild alkaline hydrolysis [4].

The –OH group acetylation provides an efficient route for the synthesis of compounds with industrial value, for the production of medicines, flavors, fragrances, solvents, food preservers, plasticizers, and cosmetics, among others [5]. The most common and inexpensive reagent used to perform the acetylation is the acetic anhydride, an activated derivative of the acetic acid, in the homogeneous phase and in the presence of bases, such as tertiary amines [6], 4-(dimethylamino)pyridine [7], and 4-pyrrolidinopyridine [8].

A number of Brønsted and Lewis acids are known to catalyze this transformation, such as *p*-toluenesulphonic acid [9], $ZnCl_2$ [10], $TaCl_5$ [11], $Zn(ClO_4)_2$ [12] and metal triflates such as $Sc(OTf)_3$ [13], $In(OTf)_3$ [14], $Cu(OTf)_2$ [15], which have been reported to show catalytic activity toward the acetylation of alcohols with acid anhydride.

However, many of these methods have some drawbacks such as low yields, long reaction time, harsh reaction conditions, use of hazardous or expensive materials and not readily available reagents [16]. The use of homogeneous catalysts has several serious complications, such as difficulty in the separation and recovery of the catalyst, disposal of the spent catalyst, and corrosion troubles. The development of easily separable and reusable solid catalysts having high activity for the acetylation reaction is, therefore, of great practical importance [16]. Recently, the use of recyclable heterogeneous acidic solid catalysts such as $GaCl_3$ [17], silica sulfuric acid [18], SBA-15-phenylsulphonic acid [19] and heteropolyacids [20], for the acetylation of alcohols or phenol by acetic anhydride, has been reported.

From the environmental point of view, a much convenient method for performing the acetylation reaction is the use of acetic acid as acetylating agent, because only water is produced as by-product.

Another advantage is the relatively cheaper route for industrial purposes when compared to the use of anhydrides or acid chlorides. The catalysts employed in the acylation of alcohols or phenols with acetic acid include mineral acids, anhydrous sulfate and catalytic amounts of sulfuric acid, aluminum phosphate/molecular sieves, sulfated zirconia, ion exchange resins, Amberlyst-15, among others [5].

We reviewed in this chapter the performance of different recyclable solid acids synthesized in our laboratory as catalysts for the use of acetic acid as acetylating agent: 1) alcohols and phenols acetylation using borated zirconia; 2) alcohols and phenols acetylation using borated zirconia modified with ammonium metatungstate; 3) isoamyl alcohol acetylation using tungstophosphoric acid supported on zirconia; and 4) isoamyl alcohol acetylation using partially substituted Cs and K salts of tungstophosphoric acid.

2. Alcohols and Phenols Catalyzed Acetylation Reactions

2.1. Alcohols and Phenols Acetylation Using Borated Zirconia

The materials formed by zirconium oxide doped with boron are solids with acid characteristics which are employed in some organic reactions, such as anisole benzoylation [21], preparation of cyclohexanoneoxime by Beckmann rearrangement [22], transesterification of β-ketoesters using ethyl acetoacetate and different alcohols as substrates [23], and C-methylation of phenol using methanol [24].

Recently, our research group prepared different catalysts based on borated zirconia, obtained by varying preparation conditions (precursor concentration used for the oxide synthesis, boron precursor concentration, and calcination temperature), and were used in the acetylation of diverse alcohols. The yield to acetylated products obtained with different alcohols (primary, benzylic and phenol) using acetic acid as acylating agent, toluene as reaction solvent, and reflux temperature was correlated with the properties of the materials used as catalyst [5]. Figure 1 shows the acetylation scheme.

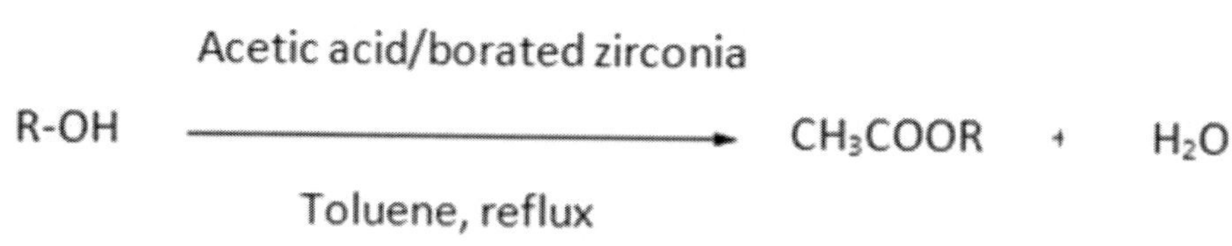

Figure 1. Alcohol acetylation using acetic acid as acylating agent.

The catalysts used in this reaction are named SXBY, where X is the number identifying the support and Y is the boron concentration (See Appendix).

The textural characteristics of the catalysts showed different behavior depending on the boron concentration. The specific surface area (S_{BET}) and the pore volume (V_p) decreased for the highest boron content (Table 1), while the mean pore size (MPS) did not vary.

Table 1. Textural properties of the catalysts

Sample	$ZrOCl_2$ concentration (10^{-2} mmolZr/cm^3)	S_{BET} (m^2/g)	V_p[a]	MPS[b] (nm)
S2B2.5	6	200.1	0.10	2.3
S2B15	6	244.5	0.14	2.3
S2B35	6	92.1	0.05	2.2
S4B2.5	5.6	125.1	0.05	2.2
S4B15	5.6	185.1	0.10	2.1
S4B35	5.6	34.1	0.02	2.5

[a]Single point total pore volume at $p/p_o = 0.98$.
[b]Mean pore size calculated by the BJH formula from the desorption branch.

It can be assumed that boron dispersed on the surface of hydrated zirconium oxide separates the particles and inhibits their later growth. The S2B35 and S4B35 solids have a proportion of boron not interacting with the support, which inhibits particle growth to a lesser extent, and the boric oxide particles can block support pores, causing a detrimental effect both on S_{BET} and on V_p.

The X-ray diffraction patterns of the catalysts calcined at 320 ºC were equal to those of the respective supports. The materials have amorphous characteristics, showing a wide band centered at 2θ value of 30° in the patterns. No line indicating the presence of crystalline phases was observed. It is worth pointing out that the calcination of bulk boric acid up to 320 ºC showed a boric oxide pattern by XRD; nevertheless, this oxide was not detected in the samples by this technique. This fact may be attributed to a good dispersion on the support as a noncrystalline phase or to the presence of crystallites small enough to give diffraction lines. On the other hand, the S4B35 sample showed an amorphous nature when calcined at 500 ºC, in agreement with the findings reported by Malshe et al. [24] for a sample with 30 mol% B_2O_3.

The FT-IR spectra of the catalysts prepared from both supports presented similar behavior. They showed a wide band extended between 400 and 1700–2000 cm^{-1} where, in turn, there are some wide bands and maxima. These complex spectra arise from the overlapping of bands corresponding to boron species and to the support. The spectra of the S2B2.5, S4B2.5, S2B15 and S4B15 samples presented bands with maxima at 1620 and 1320 cm^{-1}, assigned to bending vibrations of –(H–O–H)– and –(O–H–O)– bonds of the support. The last band appears as a shoulder in the support spectrum, but in the catalysts it is a wide band extended between 1120 and 1550 cm^{-1}, which could be attributed to tricoordinated boron bonded to the hydrated oxide structure. In addition, bands due to boron species were observed under 1100 cm^{-1}, whose assignment is difficult because, though boron in tetrahedral coordination has strong absorption in such zone, the tri-coordinated species also present bands in the same wavenumber range. In the S2B35 and S4B35 samples, it is not possible to visualize support bands, but two small peaks at 881 and 1190 cm^{-1} were observed, a position that coincides with bands of boric acid calcined at 320 ºC, whose spectrum is equal to that of a commercial B_2O_3. So, it can be assumed that small B_2O_3 aggregates are present in the samples with high boron concentration, in addition to boron species interacting with the support.

According to DTA studies the presence of boron preserved the hydrated zirconia oxide particles, increasing their thermal stability. A similar behaviour was reported for zirconia with addition of sulfate or chromium ions [25], for samples of Mo or Cu oxides on zirconia [26], for zirconia doped with Ni or Al [27], and the stability increase was attributed to the inhibiting effect on zirconia crystallization induced by the good dispersion of this type of additives.

It has been reported that the initial electrode potential (Ei) of the curves obtained by potentiometric titration with n-butylamine indicates the maximum strength of the acid sites that the titrated solid presents. Such strength can be classified according to the following scale: Ei > 100 mV (very strong sites), 0 < Ei < 100 mV (strong sites), −100 < Ei < 0 mV (weak sites), and Ei < −100 mV (very weak sites) [28]. The behaviour of the acid strength as a function of the boron concentration is similar for the samples obtained from both supports. The samples with higher proportion of boron bonded to the structure of the hydrated oxide presented higher Ei (S2B15 and S4B15 samples), possibly due to a better dispersion. Boron on the surface of the support led to more acid solids than the parent support, since boron may increase the Lewis acidity of zirconium ions by an inductive effect, as proposed by Ravindra et al. [29] for boron on silica. In addition, Brønsted acid sites may be generated by the hydration of boron species. On the other hand, the higher Ei of the samples prepared from the S4 support can be related to the highest acid strength of the support, which in turn may be due to a lower hydroxylation degree of this material that has the lowest specific surface area (Table 2). A higher amount of zirconium ions may be exposed on the surface, as was pointed out by Hadjiivanov and Lavalley [30], who have reported that some zirconium ion sites are blocked on the hydroxylated surface, and that the most acidic Lewis sites on oxide surfaces should be created at a higher dehydroxylation degree.

With these materials, the catalytic activity in acetylation reactions was explored. The effect of the catalyst amount utilized in the acetylation reaction of 2-phenoxyethanol with acetic acid was studied by employing the S4B35 catalyst. The yields obtained when using 10, 25, 50 and 100 mg of catalyst were 5, 21, 39 and 67%, respectively. So, there was an increase in the catalytic activity when the catalyst amount increased. Although a constant yield of acetylated product was not achieved, it was decided not to increase the catalyst amount and to use 100 mg in the following runs, because this value is rather high for the samples with higher boron content, i.e., 0.06 mmol B/mmol initial alcohol.

The experiments to observe the effect of boron concentration were firstly carried out with the borated zirconia samples prepared from the S4 support. The yield obtained in the acetylation of 2-phenoxyethanol with acetic acid is shown in Table 3. The values achieved were 45, 80 and 67 % for the S4B2.5, S4B15 and S4B35 catalysts, respectively.

So, the increase in the boron amount added to the support leads to an increase in the yield of acetylated product up to 15 g B_2O_3/100 g support, and then it decreases. The observed behavior is similar to that obtained for the acid strength of the samples, measured by potentiometric titration. It is worth pointing out that the yield obtained with commercial B_2O_3 was low (5%), possibly due to the low surface area of the bulk compound (S_{BET} = 5 m^2/g). The results obtained in similar tests carried out with catalysts prepared from the S2 support are also shown in Table 3. Similarly to the above-mentioned catalysts, an increase of boron from 2.5 up to 15 g B_2O_3/100 g support raised the acetylated product yield from 40 to 51%, which then decreased to 32% when the catalyst containing 35 g B_2O_3/100 g support was used.

Table 2. Acid properties of the catalysts

Sample	$ZrOCl_2$ concentration (10^{-2} mmol Zr/cm^3)	E_i (mV)
S2B2.5	6	184
S2B15	6	435
S2B35	6	415
S4B2.5	5.6	289
S4B15	5.6	522
S4B35	5.6	480

The comparison between the results obtained with both series of catalysts allows us to observe the influence of the zirconium precursor concentration employed to prepare the supports. A higher yield was obtained with the catalysts obtained using the S4 support, in accordance with their higher acid strength. As mentioned above, this is related to the characteristics of the starting support, particularly a lower hydroxylation degree.

The reuse of the catalyst is an important aspect of any industrial process. So, the reusability of the S2B15 and S4B15 catalysts was tested by carrying out repeated runs of the reaction, keeping the reactant molar ratio and temperature constant. After performing the first reaction, the catalyst is separated by filtration, dried in a vacuum system at room temperature, and used in a next reaction cycle carried out under the same previous conditions. After three cycles under reaction, both catalysts keep their activity, only a decrease of 2% in the yield was measured, so they can be recycled and reused.

When the 2-phenoxyethanol acetylation yields using the S4B35 catalyst and the same catalyst calcined at 500 °C are compared, a decrease in the catalytic activity was observed for the highest calcination temperature, as the yield decreases from 67 to 19 %. This fact may be possibly due to an increase in the size of the small B_2O_3 aggregates present in this sample when the calcination temperature was increased. When calcined at 750 °C, no reaction was detected, which may be related to the structural change toward a crystalline phase of the zirconia inhibits boron interaction, and also a decrease in S_{BET} took place (11 m^2/g).

The acetylation reaction is considered to be catalyzed by Brønsted acid sites [31] or by Lewis acid sites (metal ions in low coordination) [32]. Other authors have reported that both the Brønsted and Lewis acid sites are responsible for catalyzing the acetylation reaction [33]. Due to the characteristics of the catalysts prepared here, they may present Brønsted and Lewis sites, so we assume that both types of sites synchronously affect the acetylation activity.

The same reaction conditions were used to observe the behavior of a sample of borated zirconia, calcined at 320 °C, as catalyst in the acetylation of other alcohols and phenol with acetic acid (Table 4). The reactivity of the primary alcohols 2-phenoxyethanol and 2-phenylethanol (entries 1 and 2) was higher compared to that of the benzylic alcohol and 4-methylbenzylic alcohol (entries 3 and 4). They achieved a yield of acetylated product (67–69%), which was nearly twice that of the benzylic alcohols (25–35%). The yield difference between primary and benzylic alcohols can be attributed to steric effects that considerably affect the acetylation rate considering a Fisher esterification mechanism [34]. The presence of bulky groups, not far from the reaction center, either in the alcohol or in the acid, slows down the esterification rate, as was already reported [5].

Table 3. Influence of the boron content of catalysts obtained using the S2 and S4 supports on the acetylation of 2-phenoxyethanol with acetic acid

Catalyst	B amount (g B_2O_3/100g)	Yield (%)
S2B2.5	6	40
S2B15	6	51
S2B35	6	29
S4B2.5	5.6	45
S4B15	5.6	80
S4B35	5.6	67

Reaction conditions: 2-phenoxyethanol (1 mmol), acetic acid (1 mmol), solvent (toluene, 5 cm^3), catalyst (100 mg), temperature (110 ºC), time (14 h).

The acetylation of a sterically hindered tertiary alcohol (triphenylmethanol) was also performed, in the same conditions as 2-phenoxyethanol reaction, and only traces of acetylated product were detected. Similar differences in the behavior of primary and benzylic alcohols in the acetylation with acetic acid were reported by Mirkhani et al. [35], using a cerium polyoxometallate as catalyst.

Meanwhile, the yield in the phenol acetylation (Table 4, entry 5) was still lower than that of benzylic alcohols, reaching a yield of 10%. This behavior can be attributed to its low nucleophilicity, due to the electronic delocalization by resonance in the benzene ring. An additional experiment with 4-nitrophenol, which is less nucleophilic due to the electron-withdrawing group -NO_2, in identical conditions to phenol acetylation, only gave traces of the acetylated product. So, the reactivity order toward acetylation with acetic acid with borated zirconia was as follows: primary alcohols > benzylic alcohols > phenol. It should be noted that in all the tests performed, the selectivity was high because after the isolation and purification by column, all the unreacted starting alcohol or phenol was practically recovered.

In sum, the behavior of borated zirconia, obtained using different preparation conditions, as catalyst in the acetylation reaction of alcohols and phenol with acetic acid was presented. A good agreement between the results obtained in the acetylation of 2-phenoxyethanol and the catalyst acidity was observed. The most appropriate catalyst was borated zirconia calcined at 320 °C containing 15 g B_2O_3/100 g of a support obtained using a high concentration of $ZrOCl_2$ precursor in the starting solution, achieving a yield of 80% in acetylated product in 14 h under reaction. Besides, the catalysts kept their activity after three catalytic cycles. Though the yield was lower for benzylic alcohols and phenol, the methodology employed is a clean alternative for the acetylation reaction, tending toward eco-efficiency due to the use of relatively cheap catalysts, which can easily be prepared in the laboratory and stored for an extended period of time.

In addition, water is obtained as by-product of the reaction when acetic acid is employed as acylating agent in replacement of the classical agents.

Table 4. Acetylation of different alcohols and phenol using borated zirconia as catalyst[a]

4Entry	Alcohol	Product	Acetylated product yield (%)
1	2-Phenoxyethanol	2-Phenoxyethanol acetate	67
2	2-Phenylethanol	2-Phenylethanol acetate	69
3	Benzyl alcohol	Benzyl alcohol acetate	27
4	4-Methylbenzyl alcohol	4-Methylbenzyl alcohol acetate	25
5	Phenol	Phenol acetate	10

[a]Solvent: toluene, reaction temperature and time 100 ºC, 14 h, S4B35 catalyst mass: 100 mg.

2.2. Alcohols and Phenols Acetylation Using Borated Zirconia Modified with Ammonium Metatungstate

An attractive material to be used as catalyst support is zirconia (zirconium oxide) due to its characteristics, such as high thermal stability. The addition of different oxoanions can change the acid properties of this material [36], e.g. the sulfated zirconia presents strong acid characteristics, but the loss of sulfate species during thermal treatments or catalyst regeneration is a disadvantage [37].

On the other hand, a widely studied catalyst is tungstated zirconia [38–40], as an environmentally friendly alternative to sulfated zirconia [41]. Tungsten-modified zirconia has been tested as catalyst in many reactions, such as tert-butylation [42], esterification [43], as well as in many major industrial applications [44]. So, it may be interesting to study the effect of the addition of a tungstate anion to borated zirconia [45]. The acetylation reaction is very important due to the wide spectrum of applications (see e.g. [46–48]), in particular for the protection of OH groups [49], a reaction that is employed to obtain compounds of interest for the pharmaceutical, cosmetics or food industries, among others [50], as it was above-mentioned. In this context, borated zirconia modified with a tungstate anion was prepared in order to observe the effect of this addition on both the textural, physical-chemical and acid characteristics of the materials and the behavior as catalyst in alcohol acetylation, with acetic acid instead of the classical acetic anhydride as acetylating agent. Two levels of tungsten concentration, from ammonium metatungstate (AMT), were added to two borated zirconias, prepared using different precursor concentrations (see Appendix). The unique boron concentration used was selected based on previous results [5]. For comparative purposes, only tungsten was added to the same bare zirconias. The nomenclature of the catalysts is given in the Appendix.

The catalysts presented a mean pore size, estimated by the BJH formula from the desorption branch of the isotherm, higher than 2 nm, the lowest limit of mesoporous size according to IUPAC classification. The samples containing only boron, S2B15 and S4B15, have a specific surface area (S_{BET}) higher than that of the respective parent supports (214 and 146 m^2/g respectively). This fact may be explained considering that boron is well dispersed on the surface of the hydrated zirconium oxide and inhibits particle growth [26]. Nevertheless, the specific surface area of the S2B15W12.5, S2B15W25, S4B15W12.5 and S4B15W25 samples, with tungsten addition to borated zirconia, decreased between 20-40 % with respect to the value of the respective borated zirconia (Table 5).

For these samples, the micropore area estimated from the t-plot analysis is low, it is 6–25% of the corresponding S_{BET} value. The total pore volume, determined as the single point at p/po = 0.98, remained almost constant in a relatively high value (0.10 cm^3/g), and the mean pore size increased to 2.7-3.4 nm compared to the corresponding borated zirconia (2.1-2.3 nm). In turn, the textural characteristics of the zirconia samples containing tungsten, prepared by directly impregnating zirconia, were different to those obtained by modifying the borated zirconia. The first samples have a specific surface area markedly lower than that of the corresponding support.

Table 5. Specific surface areas of the prepared catalysts

Catalyst	Specific surface area (m^2/g)[a]	Micropore area (m^2/g)[b]
S2B15	245	25
S2B15W12.5	182	11
S2B15W25	144	13
S2W12.5	74	7
S2W25	94	9
S4B15	185	60
S4B15W12.5	113	29
S4B15W25	149	33
S4W12.5	34	-
S4W25	27	-

[a]Estimated by the BET method.
[b]Estimated from t-plot analysis.

Thus, the S2W12.5 and S2W25 samples presented a specific surface area 60% lower than that of the S2 support, and S_{BET} notably decreased for the S4W12.5 and S4W25 solids, being around 80% lower than that of the parent S4 zirconia (Table 5). This fact indicates that the preparation method used, involving an impregnation in excess of solution followed by evaporation of the excess solution, may lead to tungsten species that are not well dispersed on the support, as may be inferred from XRD results.

So, the tungsten species do not help in separating the particles that produces the inhibition of their later growth, as seems to be the process that occurs in the presence of boron. This effect is accompanied by a decrease of the pore volume (0.02-0.06 cm^3/g) and an increase of the mean pore size (2.6-3.5 nm), possibly due to a blocking of the smaller pores. So, it was observed that the modification of zirconia with boron was more efficient than the modification with tungsten when referring to the preservation of the textural properties of the materials, a behavior that can be attributed to a better dispersion of boron species than tungsten species on the support.

The X-ray diffraction patterns of the S4B15W12.5 and S4B15W25 solids were similar to that of S4B15 borated zirconia, and the same behavior was observed for the S2B15W12.5 and S2B15W25 samples, compared to S2B15. All the patterns presented a wide band centered at a 2θ value of 30°, indicative of solids with mainly amorphous characteristics. Besides, lines attributable to boric oxide or to a tungsten oxide produced by a partial decomposition of metatungstate were not detected in the samples.

This fact may be attributed to a good dispersion of the boron and tungsten species on the support or to crystallites small enough to give diffraction lines. However, in the S4W25 sample pattern, lines overlapped to such band in the 23–24 °2θ zone, where the WO_3 presents the lines of higher intensity, were visualized and a peak at a 2θ value of 9.5° appears, which coincides with the highest peak of ammonium metatungstate without thermal treatment.

By DTA analysis, the diagrams of all the samples presented an endothermic peak with a maximum at 40–50 °C, which can be assigned to a loss of physisorbed water. The borated zirconias with tungsten addition showed an exothermic peak placed at 720–730 °C in the DTA diagrams, which is assigned to the transformation of the amorphous hydrated oxide into a crystalline phase.

The monoclinic phase of zirconia is stable up to 1200 °C, though a metastable tetragonal phase can appear up to 650 °C, due to the effect of impurities or to the crystallite size [26]. The exothermic peak appeared at a higher temperature than for the corresponding S2 and S4 parent supports, and at a slightly lower temperature than for the corresponding borated zirconia. The presence and the interaction of boron on the support preserves the microparticles of the hydrated oxide [25], thus increasing its thermal stability; however, the temperature at which the exothermic peak maxima appeared slightly decreased with tungsten addition. The increase of stability was attributed to the inhibiting effect on zirconia crystallization that induces a good dispersion of the additives added to the oxide [27, 33]. However, the behavior is different when zirconia is modified only with tungsten. While the parent support showed only one exothermic peak, the diagrams of the solids containing 12.5 or 25 %W showed three wide exothermic peaks. The diagrams of the samples have similar features to those of AMT. This fact may be indicative of the presence of similar species, and the shifting of the maxima towards higher temperatures can be attributed to interaction with the support.

It can be remarked that the XRD pattern of AMT calcined at 320 °C was similar to the pattern of a tungsten and ammonium oxide (JCPDS 42-0452), which can be the product of a partial decomposition of ammonium metatungstate. When AMT was calcined at 450, 550 or 750 °C, WO_3 appeared, so the bands with maxima at 520 and 658 °C observed in DTA diagrams may be a consequence of structural changes of WO_3.

The Ei values of the catalysts determined using potentiometric tritation are presented in Table 6. When the samples containing boron are considered, the addition of tungsten led to a decrease of the maximum acid strength compared with the parent borated zirconia. The higher value of Ei observed for the samples prepared from the S4 support can be related to the higher initial potential value obtained for the parent support (Table 6). Ei values were 389 and 438 mV for S2 and S4 supports, respectively.

This behavior was attributed to a lower hydroxylation degree of the S4 solid that has a lower specific surface area than that of the S2 support as was previously explained [30]. In turn, the acid strength increased when only tungsten was added to zirconia, significantly in the samples obtained from S2 zirconia.

So, from the characterization results, it can be summarized that tungsten-modified borated zirconias have characteristics that depend on the preparation method used, the concentration of zirconium precursor employed in obtaining the parent zirconias, as they confer different textural and acidic characteristics to the two supports that were then impregnated and, to a lesser extent, on tungsten concentration. The solids were predominantly amorphous and mesoporous, presenting very strong acidity. The addition of boron increased the range of stability temperature of the hydrated oxide, as a result of good dispersion. This fact also led to good tungsten dispersion, when the behavior is compared to that of samples in which only tungsten is added to zirconia.

Table 6. Acid properties of the catalysts

Sample	Tungsten concentration (g W/100 g support)	E_i (mV)
S2B15	0	435
S2B15W12.5	12.5	333
S2B15W25	25	382
S4B15	0	522
S4B15W12.5	12.5	463
S4B15W25	25	447
S2	0	389
S2W12.5	12.5	515
S2W25	25	578
S4	0	438
S4W12.5	12.5	420
S4W25	25	490

The acetylation of 2-phenoxyethanol with acetic acid using borated zirconias modified with ammonium metatungstate led to high yield values, as can be observed in Table 7.

The catalysts obtained from the parent S4 zirconia gave yields to acetylated product higher than the one corresponding to the S2 support, in accordance with their higher acid strength, as was already observed for borated zirconias with different boron content without the addition of tungsten. The modification with tungsten only led to a higher yield for the S4B15W12.5 sample, showing that the expected effect is not efficiently achieved. On the other hand, the acetylation of 2-phenoxyethanol with acetic acid using tungsten modified zirconia as catalyst did not lead to high yields to ethyl 2-phenoxyethanoate. The obtained values were 25%, 30%, 20% and 21% for the S2W12.5, S2W25, S4W12.5 and S4W25 samples, respectively. This was not expected because the acid strength determined for these samples is high, but evidently the low surface area and poor tungsten dispersion, reflected in the lines observed by XRD, have an adverse effect.

If we considered each group of samples containing tungsten, with and without boron (Table 8) individually, a direct relation to the acidity was found for the yield to acetylated product, which may be dependent on the preparation conditions of the materials. It is clearly observed that the concentration of the zirconium oxide precursor used and the boron addition to zirconia previous to the modification with tungsten are important variables, tungsten addition and tungsten concentration being less important. The properties acquired by these materials, mainly regarding the surface area, acidity and dispersion, are important for their behavior as acetylation catalysts. In order to investigate the reusable properties of the tungsten-modified borated zirconia, experiments were carried out using the S4B15W12.5 catalyst. This process was repeated twice without important loss of catalyst activity. The corresponding acetate was obtained in yields of 89, 87, and 87% for successive reuses.

Table 7. Yield to acetylated product obtained with S2B15 and S4B15 borated zirconias, and S2B15W12.5, S2B15W25, S4B15W12.5 and S4B15W25 tungsten-modified samples

Entry	Catalyst	W (%)	Yield (%)
1	S4B15W	0	80
2		12.5	83
3		25	78
4	S2B15W	0	51
5		12.5	40
6		25	50

Reaction conditions: 2-phenoxyethanol (1 mmol), acetic acid (1 mmol), solvent (toluene, 5 cm^3), catalyst (100 mg), temperature (110 ºC), time (14 h).

Table 8. Relation between the maximum acid strength and the yield to acetylated product for samples with and without boron

Sample	E_i (mV)	Yield (%)
S2B15W12.5	333	40
S2B15W25	382	51
S4B15W12.5	463	87
S4B15W25	447	72
S2W12.5	515	25
S2W25	578	30
S4W12.5	420	20
S4W25	490	21

The potential of this catalytic system for the acetylation of various alcohols (primary, secondary, tertiary and benzylic), and phenols with electron-donating and electron-withdrawing groups was also explored. The reactivity of the selected compounds was tested using the same reaction conditions and S4B15W12.5 as catalyst, which is the solid that gave the highest conversion in 2-phenoxyethanol acetylation. Results of the obtained yields are listed in Table 9. It must be pointed out that the reactions were clean and the products were isolated by liquid column chromatography in pure form without further purification. The reaction is very selective and no competitive side reactions take place, as the gas-chromatography/mass spectrometry analysis showed. However, an important effect of the alcohol structure on the yield to the reaction products was found.

The observed reactivity trend for alcohols and phenols can be conceptualized using the catalysis model for esterification. Initially, two components contribute to the alcohol or phenol reactivity: an inductive effect and a steric effect. The inductive effect is dependent on nucleophilicity of the alcohol. The steric component affecting alcohol reactivity is also a decisive factor. Steric hindrance increases with molecular size, inducing electronic repulsion between nonbonded atoms of reacting molecules. This repulsive hindrance decreases the electronic density in the intermolecular region and disturbs the bonding interaction.

The reactivity of 2-phenylethanol and 3-phenylpropanol primary alcohols, and 4-methylbenzyl alcohol, a benzylic alcohol (Table 9, entries 1–3), was higher than that of 1-phenylethanol, a secondary alcohol (Table 9, entry 4). The yield difference between primary and secondary alcohols can be attributed to steric effects that considerably affect the acetylation rate. The presence of bulky groups, not far from the reaction center, either in the alcohol or in the acid, slows down the esterification rate, in our case, where the most sterically hindered alcohols give a lower yield. Similarly, very bulky menthol (a secondary alcohol) and triphenylmethanol (a tertiary alcohol) give only traces or no reaction is detected under similar reaction conditions (Table 9, entries 5 and 6).

On the other hand, the yield in phenol acetylation is still lower than that of the primary and secondary alcohols, reaching only yields of 10–15% for different phenols (Table 9, entries 7–9). This behavior can be attributed to its low nucleophilicity, due to the electronic delocalization by resonance in the benzene ring. 4-Nitrophenol, a phenol with a strong electron-withdrawing group, fails to give the expected acetylation product (Table 9, entry 10). In sum, the reactivity order toward acetylation with acetic acid was as follows: primary alcohols > secondary alcohols > phenols.

Table 9. Acetylation of different alcohols and phenol using S4B15W12.5 catalyst[a]

Entry	Substrate	Product	Yield (%)
1	2-Phenylethanol	2-Phenyletanol acetate	83
2	3-Phenylpropanol	3-Phenylpropanol acetate	75
3	4-Methylbenzyl alcohol	4-Methylbenzyl alcohol acetate	71
4	α-Methylbenzyl alcohol	α-Methylbenzyl alcohol acetate	62
5	Menthol	Menthol acetate	traces
6	Triphenyl methanol	Triphenyl methanol acetate	-
7	Phenol	Phenol acetate	15
8	4-Methoxyphenol	4-Methoxyphenol acetate	15
9	4-Chlorophenol	4-Chlorophenol acetate	10
10	4-Nitrophenol	4-Nitrophenol acetate	traces

[a]Reaction conditions: alcohol (1 mmol), acetic acid (5 mmol), toluene (5 cm^3), S4B15W12.5 catalyst mass (100 mg), reaction temperature and time (110 ºC, 14 h).

In short, the acetylation of 2-phenoxyethanol with acetic acid employing borated zirconias modified with tungsten led to good yields to acetylated product, unlike tungstated zirconias that gave low yields. The better yields were reached with the catalysts obtained using borated zirconia prepared with a high concentration of zirconium precursor as support. The alcohol and phenol structure is a key in the yield to the acetylation product.

2.3. Isoamyl Alcohol Acetylation Using Tungstophosphoric Acid Supported on Zirconia

Catalysis by heteropolyacids (HPA) and related compounds is a field of increasing importance. The HPA with Keggin-type structure [51] possess special characteristics that make them very useful in catalysis [52]. These HPA are stronger acids and they have a significantly higher catalytic activity in homogeneous and heterogeneous systems compared with conventional acid catalysts as sulphuric acid, silica–alumina or ion-exchange resins, among others.

The use of HPA as catalysts is important in the development of clean technologies, since it avoids the drawbacks of environmental pollution and corrosion of the conventional technologies. Other advantages that can be emphasized are the ability of recovering and reusing them in liquid phase reactions compared to the homogeneous catalysts and the possibility of their use in continuous processes.

There are many acid reactions in which this kind of catalysts can be used, as dehydration, cyclization, alkylation, condensation, esterification and etherification [53, 54], among others. A disadvantage of bulk HPA as catalysts lies in their relatively low stability and, also, their low surface area. To minimize these disadvantages, the HPA are usually supported on a carrier. Acidic or neutral solids such as active carbon, acidic ion-exchange resin, SiO_2 and ZrO_2 are suitable as supports [54–59]. Supporting heteropolyacids on solids with high surface area is also a useful method for improving catalytic performance. However, it is necessary to consider carefully the changes in the acid strength, the structure of the aggregates and the possibility of decomposition.

The study of ZrO_2 is of interest due to its already mentioned properties, which allow their use as catalyst or support. The catalytic behavior of ZrO_2 in acidic reactions depends on the preparation method, the thermal treatments and, the addition of certain modifiers. So, the addition of sulphate or phosphate, among others, to freshly precipitated zirconia has yielded solid materials with strong acidity and catalytic properties, greater than those of the single components [26, 41, 60–62].

The nature of the support and HPA loading may have an important effect on the activity and stability of the catalysts, particularly due to the characteristics of the interaction between the HPA and the ZrO_2 support [63]. These supports can be prepared by means of different gelation agents and temperatures of pre-treatments.

HPA/ZrO_2 solids based on tungstophosphoric acid (TPA) were prepared [64], supporting this heteropolyacid from their water–ethanol solutions on zirconia (Z), both by the equilibrium adsorption (E) and the incipient wetness (W) impregnation methods, and used as catalysts in the acetylation of isoamyl alcohol with acetic acid to obtain isoamyl acetate (banana oil).

In the preparation of supported HPA as acid catalysts, the identification of the species present in the impregnating solutions is particularly important. In aqueous solutions of TPA, the $[PW_{12}O_{40}]^{3-}$anion have a limited stability range; at pH 1.5–2, they are reversibly and quickly transformed into the lacunar species $[PW_{11}O_{39}]^{7-}$ [60]. Thus, owing to the restricted stability in aqueous solutions and bearing in mind previous reports indicating that the addition of an organic solvent has a stabilizing effect on the Keggin type heteropolyanions in solution [65], TPA solutions in e-w 50% (v/v) were used.

The UV-visible spectra of HPA solutions in e-w before the contact with the support show a band at 265 nm for TPA, in agreement with Nomiya et al. [66]. This band may be assigned to the oxygen–metal charge-transfer band of the tungstophosphate anions. The UV-visible spectra of the initial TPA solution and those obtained at different impregnation times of the Z support with the solution of 120 g W/dm^3 showed only the characteristic band of the $[PW_{12}O_{40}]^{3-}$anion. A decrease of the band intensity with time was observed, due to the decrease of the solution concentration while the anion was adsorbed on the support surface. So, the stabilizing effect of ethanol–water solution as solvent compared to water [67] for preparing the impregnation solutions was verified.

The Z support has mainly a microporous structure, with pore diameters less than 4 nm and specific surface area of 214 m^2/g. Each Keggin anion can be inscribed into a sphere of center X and radius approximate to 0.52 nm (radius = mean of the 12 X–Od distances, where Od is a terminal oxygen atom and X = P in our case) [68]. So, the penetration, and later adsorption, of the Keggin anions into the pores of the support could be sterically affected, as it was observed in a previous paper [63].

The S_{BET} of Z-TPA-W catalysts are lower than that of the support (Table 10). On the other hand, the S_{BET} also decreases when Z support is impregnated using the equilibrium adsorption method, but in this case the decrease is higher. This may be explained taking into account the different preparation methods employed to obtain the catalysts. In the incipient wetness method, the solution transport into the pores of the support is only performed by capillary action.

On the other hand, in the equilibrium adsorption impregnation technique, the solute transport is the result of convective flow till the pores are filled up and then by diffusive mechanism, if a solute–support interaction exists.

Table 10. Total HPA concentration and specific surface area of support and catalysts

Catalyst	C_T (mmol TPA/100 g cat)	S_{BET} (m^2/g)
Z support	-	214
Z-TPA-E	5.24	186.1
Z-TPA-W	2.23	210.8

This is the case for the TPA/Z system, so the higher impregnation time and impregnating solution volume involved in the equilibrium adsorption method led to an additional solute transport by diffusion. Then, the catalysts prepared by this latter technique contain a higher total TPA concentration (C_T) compared to those prepared by the other method, when the same initial concentration of the impregnating solution is used. Thus the S_{BET} decrease, which may be attributed to the partial obstruction of the support pores, is more important than in the wetness impregnation. It may be pointed out that the S_{BET} changes caused by the impregnation are lower than that observed in a previous paper [63]. This is due to the lower concentration of the impregnating solutions, and the higher mean pore diameter and pore volume of the support used in the present work. These different textural characteristics were a consequence of the higher aging and washing times employed in the support preparation.

The XRD patterns of all the prepared samples are similar to that of the support. This may be a consequence of the HPA impregnated on the support are highly dispersed as a non-crystalline form or crystallites low enough to give diffraction lines are present. The adsorption of metallic oxo anions over oxide supports has been studied by several authors who have interpreted their results in terms of the oxide IEP (isoelectric point) [69-72]. Zirconia gel has an IEP between 6 and 7, typical of amphoterous transition oxides. Chlorine-free zirconia, prepared from n-butoxide, has an IEP of 7. The zirconia prepared from chloride precursors has always lower values of IEP, due to the presence of chloride ions retained in the gel matrix [73]; in our case the IEP is 6. This characteristic of the support surface leads to a particular interaction with the HPA, depending on the operative variables of the impregnation technique employed.

The charge transfer absorption spectra of most nonreduced polyanions obtained by DRS appear in the 200–500 nm range, and consist of bands which may be ascribed to oxygen-to-metal transfers. The bulk TPA spectrum presents a band at 212 nm and another broad band that extends from 250 to 450 nm [26]. The characterization of the catalysts obtained by incipient wetness impregnation (Z-TPA-W) and equilibrium adsorption impregnation (Z-TPA-E), carried out by DRS, showed that the spectra present a band approximately at 220 nm and another band which extends up to 400 nm (Figure 2). Therefore, in all cases, bands shorter than that corresponding to the bulk acid were observed. This may be a result of a possible partial degradation of the HPA when they interact with the support.

The FT-IR spectrum of ZrO_2 presents a wide band with maximum between 400 and 700 cm^{-1} which extends up to 1150 cm^{-1}.For the bulk acid $H_3PW_{12}O_{40}$, the spectrum presents bands at 1081, 982, 888, 793, 595 and 524 cm^{-1} (Figure 3). On the other hand, the bands of the $[PW_{11}O_{39}]^{7-}$ anion are placed according to the literature at 1100, 1046, 958, 904,812 and 742 cm^{-1} [74].

The spectra of Z-TPA-E solid (Figure 3) presents bands at 1079, 981and 888 cm^{-1}, which may be assigned to bulk TPA, and also bands at 1095, 1058, 964 and 816 cm^{-1} appear in the spectra, which may be assigned to the lacunar phase $[PW_{11}O_{39}]^{7-}$ or to dimeric species [74].

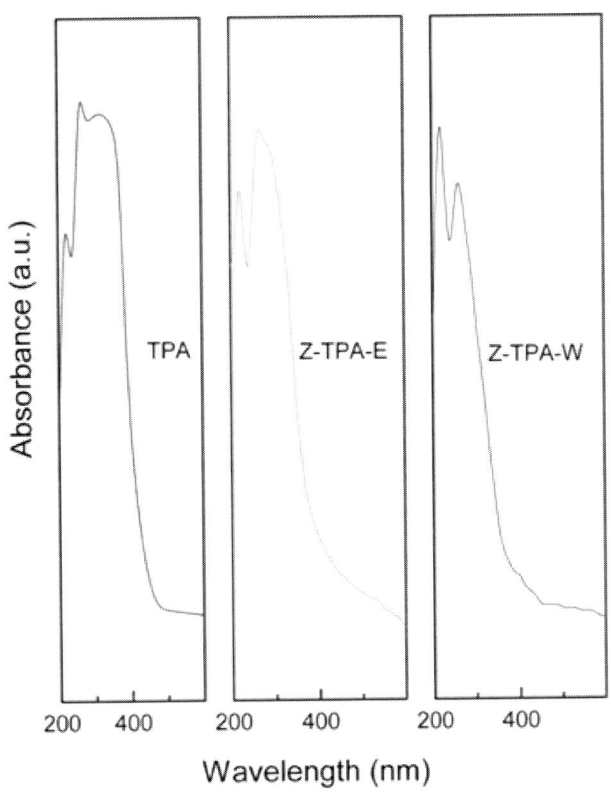

Figure 2. DRS spectra of bulk and supported TPA.

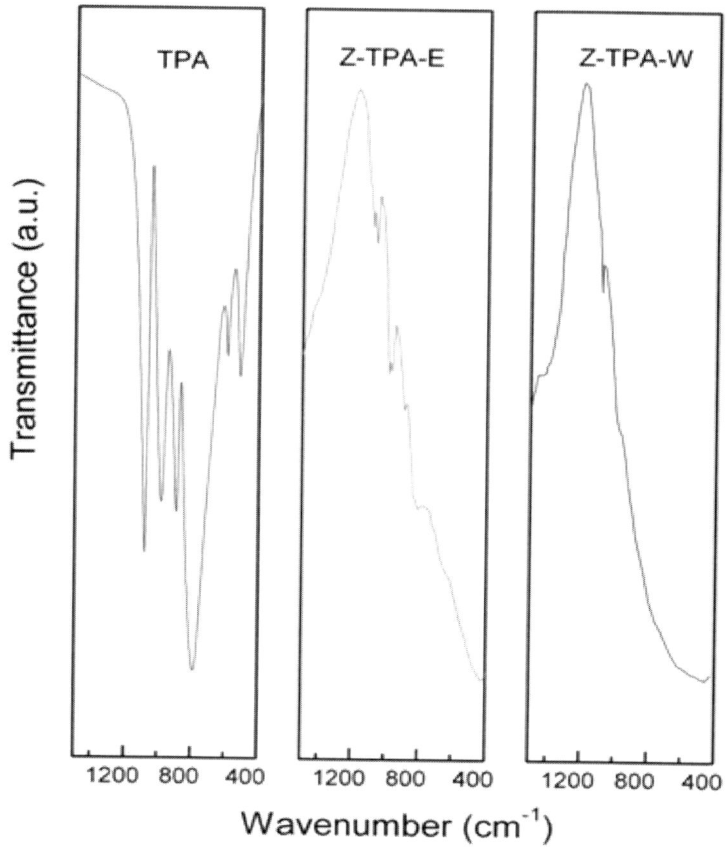

Figure 3. FT-IR spectra of bulk and supported TPA.

The partial transformation of the $[PW_{12}O_{40}]^{3-}$ anion into other species can take place as a result of the interaction with the support and/or the later drying process. The Z-TPA-W spectra shows bands of low intensity, probably due to their low HPA concentration which favours that they are masked by the wide band of the support. As a consequence, the species present in these catalysts could not be well defined.

For a better definition of the species resulting through the interaction of the heteropolyacid with the support, wetness and in equilibrium impregnated, a study by ^{31}P MAS-NMR was performed. According to the literature [56], the bulk TPA chemical shift is observed between −15 and −15.6 ppm, depending on the hydration level. On the other hand, the chemical shift of the dimeric anion $[P_2W_{21}O_{71}]^{6-}$ was reported at −13.3 ppm [75], and that of the lacunar anions $[PW_{11}O_{39}]^{7-}$ at −10.4 ppm [76].

The recorded spectrum of the Z-TPA-E catalyst shows (Figure 4) a partial transformation of the $[PW_{12}O_{40}]^{3-}$ anion (−14.8 ppm) into the dimeric $[P_2W_{21}O_{71}]^{6-}$ species (−13.3 ppm). On the other hand, for the Z-TPA-W catalyst, obtained by the incipient wetness impregnation technique, the recorded spectrum shows the chemical shift corresponding to the bulk acid in slight interaction or without interaction with the support (−14.7 ppm) [26, 74].

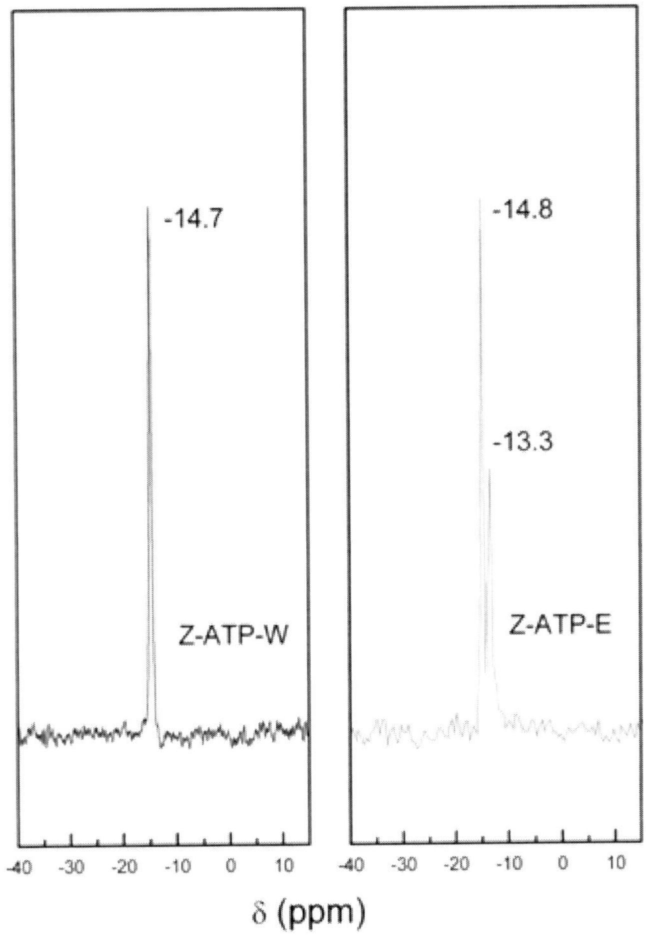

Figure 4. ^{31}P MAS-NMR of supported TPA.

Then, the study of the catalysts by NMR allowed us to verify that the species present in the solid coincided with those in the impregnating solution in contact with the support only when the wetness impregnation method was used.

During the equilibrium impregnation, the interaction degree between the $[PW_{12}O_{40}]^{3-}$ anions and the surface hydroxyl groups of the support is higher than in the above mentioned solid, which leads to a partial degradation of the Keggin anion into dimeric species.

The acidity which presents the ZrO_2, which is a weak acidic oxide, is modified by the subsequent impregnation, fundamentally due to the interaction of the TPA anion with the support surface groups. The titration curves obtained for bulk TPA, zirconia and the different supported catalysts allow us to estimate the acidity of the different solids prepared (Figure 5).

The catalysts obtained by impregnation of TPA on Z have a higher acidic strength. In the case of Z-TPA-E, the initial potential value is 501 mV and it is 402 mV for Z-TPA-W, both slightly lower than that corresponding to the bulk acid (550 mV). The number of acid sites is very similar for both catalysts.

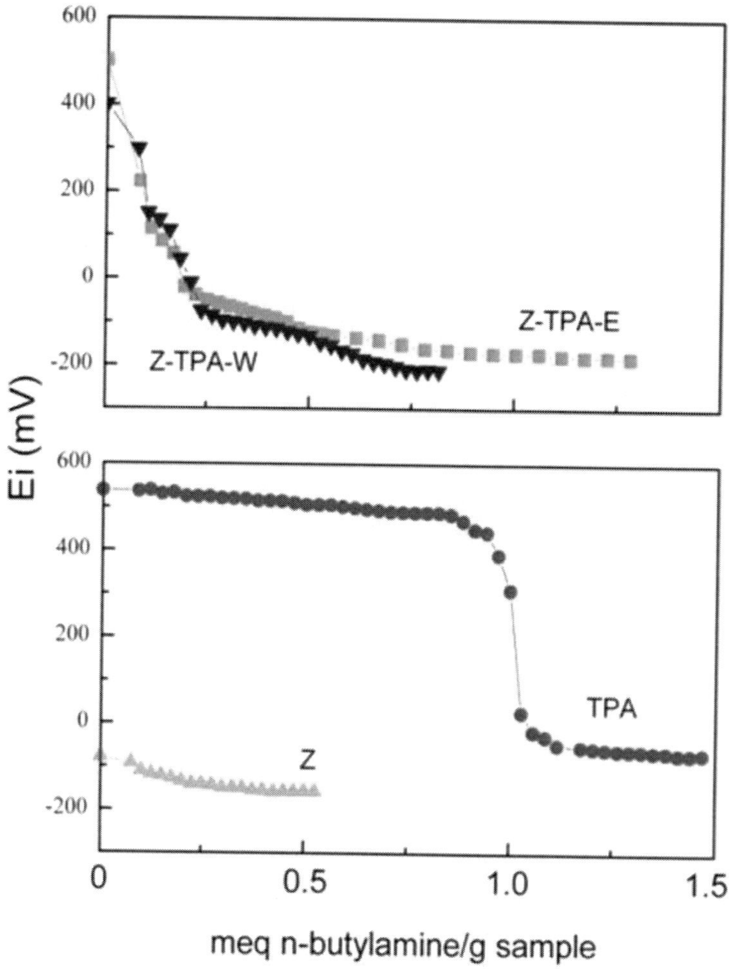

Figure 5. Potentiometric titration of bulk and supported TPA.

On the other hand, the TPA adsorbed concentration (Ca), determined by an equilibrium adsorption isotherm, is low (3.78 mmol TPA/100 g cat), indicating a low TPA–support interaction, which explains the higher amount of free protons and so the higher acidity observed. Nevertheless, the solids present a higher acidity than that corresponding to the support, −80 mV, which indicates that it is a weak acid oxide.

The use of the zirconia-supported TPA catalysts in the acetylation of isoamyl alcohol with acetic acid showed high selectivity to the ester formation, though only traces of 3-methyl-1-butene, 2-methyl-1-butene and isovaleric aldehyde were detected by means of GCMS.

The values of turnover number (TON) for the Z-supported catalysts are shown in Table 11. It was observed that the catalytic activity of these supported HPA catalysts for the acetylation is higher than that of bulk HPA. This could be explained taking into account that the catalytically active surface increases when the acid is dispersed on the support.

For this reaction, the catalyst prepared by the incipient wetness method, Z-TPA-W, show higher TON values than the corresponding catalyst prepared by the equilibrium method, Z-TPA-E. This behaviour is mainly due to the lower species dispersion on the surface of the catalysts prepared by the equilibrium method. This is a consequence of the higher C_T in these catalysts compared with the corresponding to the catalysts obtained by the incipient wetness method, as above mentioned. Another possible reason is that the species present on these last catalysts are the Keggin anions while in the catalysts prepared by the equilibrium method a partial degradation of Keggin anions exists.

In short, in the operative conditions of preparation used, the catalyst obtained by incipient wetness impregnation of zirconia with tungstophosphoric is the most active of the prepared catalysts, for the acetylation reaction which leads to the isoamyl acetate, thus obtaining an important product for flavor and fragrance industries, with an ecofriendly technology.

2.4. Isoamyl Alcohol Acetylation Using Partially Substituted Cs and K Salts of Tungstophosphoric Acid

For acidic heterogeneous reactions, bulk heteropolyacids generally exhibit low catalytic performance due to their low surface area. Supporting the HPA on solids with high surface area is a useful method for improving catalytic performance in liquid–solid and gas–solid surface heterogeneous reactions. Although the surface area of HPA is low, salts of large alkaline ions as Rb^+, Cs^+ and K^+ displayed high surface area and microporous structure [77]. So that, the wet impregnation of HPA on these salts can be considered as an alternative method to prepare supported HPA catalysts. This procedure preserves the Keggin structure of the acid phase, even for low HPA amounts [78].

Table 11. Turnover number in the acetylation of isoamyl alcohol with acetic acid

Entry	Catalyst	TON (mol ester/mol HPA)
1	TPA	6.48×10^2
2	Z-TPA-W	1.40×10^4
3	Z-TPA-E	7.93×10^3

On the other hand, acidic cesium salts of tungstophosphoric acid, $Cs_XH_{3-X}PW_{12}O_{40}$, prepared by precipitation in aqueous solution, have been shown as highly active catalysts in a number of acid type reactions. These salts can be considered as two phase systems where the acid phase is effectively dispersed on the neutral salt which displays a much higher surface area than the parent acid. The enhancement of the surface area results in an increase in the specific catalytic activity in spite of an overall lower number of acid sites.

In order to estimate the acidity of solid catalysts, several methods have been used. Titration with Hammett indicators, temperature programmed desorption of adsorbed molecules such as ammonia or pyridine, adsorption microcalorimetry, NMR spectroscopy and catalytic probe reactions are the most common methods for characterizing solid acids. As no single technique provides a complete certainty, the use of more than one method is recommendable to get a more accurate description of the nature of the solid acidity, thus overcoming the intrinsic restrictions of every technique. We have used the catalytic activity in the isopropanol dehydration to rank the acidity of diverse catalysts based on tungstophosphoric acid supported on carbon, silica, titania or alumina [26]. The acidity of the catalysts was found to be mainly related to the characteristics of the interaction between the solute and the support, to the species present on the carrier surface and to the support nature. In addition to this catalytic probe reaction method, we have utilized potentiometric titration with n-butylamine to estimate the acidic properties, as those of TPA immobilized in polyvinyl alcohol hydrogel beads [79] or supported on zirconia [64]. We have prepared and characterized solid acid catalysts based on $H_3PW_{12}O_{40}$ using partially substituted $Cs(K)_X H_{3-X}PW_{12}O_{40}$ salts as support [80]. The acidic properties were examined through the catalytic probe reaction of isopropanol dehydration and by potentiometric titration. We studied the synthesized solids as acetylation catalysts and made correlation between their main characteristics and the activity in the acetylation of isoamyl alcohol with acetic acid to obtain isoamyl acetate. The specific surface area (S_{BET}) of the solids determined from N_2 adsorption–desorption isotherms are shown in Table 12 together with the microporous area (S_{micro}) calculated according to the MP method using t-plot. The partially substituted salts presented high S_{BET} compared to the bulk acid as already reported by several authors (see e.g. [81-83]). Besides, the potassium salts displayed lower surface area than the cesium salts, related to the lower diameter of the potassium cation, and the same trend is shown by the less substituted salts compared to $Cs(K)_{2.9}H_{0.1}PW_{12}O_{40}$ salts. Similar behavior has been reported by Parent and Moffat [84] for the silver, thallium and cesium stoichiometric and nonstoichiometric salts of TPA. In addition, according to the calculated S_{micro} values, 90% of the total surface area came from microporous structure. Thus, the main contribution to the surface area is due to the presence of micropores formed as a consequence of the substitution of hydrated protons of TPA by cesium or potassium cations.

Table 12. Specific surface area of the cesium or potassium salts

Catalyst	S_{BET} (m^2/g)	S_{micro} (m^2/g)	S_{meso} (m^2/g)
TPA	2	-	-
$Cs_{2.9}H_{0.1}PW_{12}O_{40}$	139	122	17
$K_{2.9}H_{0.1}PW_{12}O_{40}$	125	113	12
$Cs_{2.6}H_{0.4}PW_{12}O_{40}$	121	108	13
$K_{2.5}H_{0.5}PW_{12}O_{40}$	109	99	10

The FT-IR spectra of $Cs_{2.9}H_{0.1}PW_{12}O_{40}$, $Cs_{2.6}H_{0.4}PW_{12}O_{40}$, $K_{2.9}H_{0.1}PW_{12}O_{40}$ and $K_{2.5}H_{0.5}PW_{12}O_{40}$ samples dried at 70 °C presented the same characteristic bands of bulk acid $H_3PW_{12}O_{40}$, however, a splitting of the W–O_d band into two components was observed, one maximum is placed at 985 cm^{-1} for all the samples and the other is at higher frequency (994, 997, 990 and 995 cm^{-1}, respectively). The new component may be assigned to the direct interaction between the $[PW_{12}O_{40}]^{3-}$anion and the Cs^+ or K^+ cations [85]. Similar spectra were obtained for the impregnated salts. The results shown here allow us to establish that the primary Keggin structure remained unaltered. Neither during the preparation of the salts nor during the impregnation, a detectable transformation of the $[PW_{12}O_{40}]^{3-}$ anions into the $[PW_{11}O_{39}]^{7-}$ lacunary species was not noticed.

The powder XRD patterns of $Cs_{2.9}H_{0.1}PW_{12}O_{40}$ and $Cs_{2.6}H_{0.4}PW_{12}O_{40}$, salts (Figure 6) presented the characteristic peaks corresponding to the $Cs_3PW_{12}O_{40}$ cubic phase [86], placed at the same position for both partially substituted salts.

With regard to $K_{2.9}H_{0.1}PW_{12}O_{40}$ and $K_{2.5}H_{0.5}PW_{12}O_{40}$ samples, the XRD patterns also presented the characteristic peaks of the body centered cubic structure. Nevertheless, it must be noted that the peaks are shifted to higher diffraction angles compared to cesium based samples, indicative of a smaller lattice parameter.

The ^{31}P MAS–NMR spectra of $Cs_{2.9}H_{0.1}PW_{12}O_{40}$, $Cs_{2.6}H_{0.4}PW_{12}O_{40}$, $K_{2.9}H_{0.1}PW_{12}O_{40}$ and $K_{2.5}H_{0.5}PW_{12}O_{40}$ samples, dried at 70 °C, showed only one peak at a chemical shift of −14.8, −14.7, −14.6 and −14.7 ppm, respectively. It may be attributed to $[PW_{12}O_{40}]^{3-}$ species, in good agreement with results reported for pure $Cs_3PW_{12}O_{40}$ (−14.9 ppm) [87, 88]. All the samples presented, in addition, a small shoulder at around −15.1 ppm, which may be assigned to the presence of dispersed TPA. This technique reinforced FT-IR conclusion about undegraded primary Keggin structure.

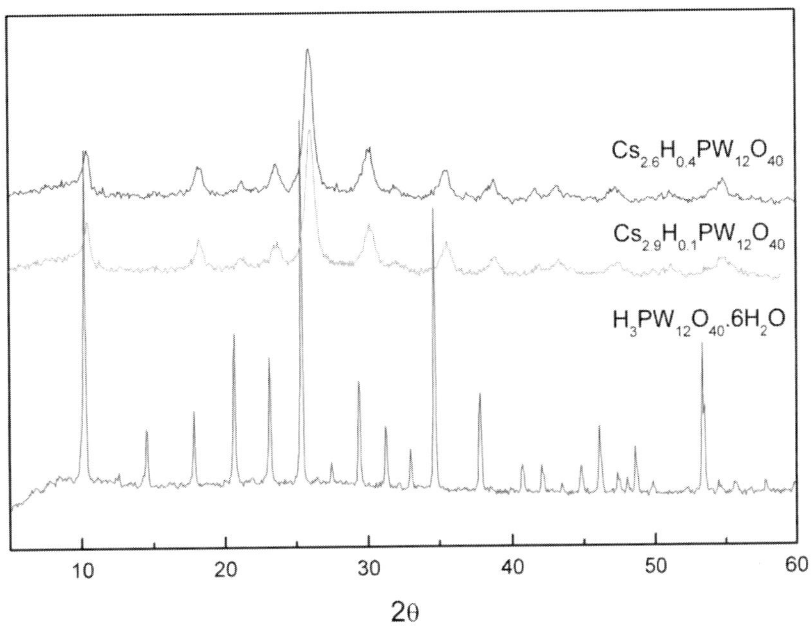

Figure 6. XRD patterns of the partially substituted cesium salts and $H_3PW_{12}O_{40}.6H_2O$.

In agreement with Essayem et al. [83], the absence of shifting for the XRD peaks with the composition and the presence of only one species by ^{31}P MAS–NMR suggest that the partially substituted samples may be considered as $Cs(K)_3PW_{12}O_{40}$ salt with a given amount of well dispersed $H_3PW_{12}O_{40}$ entrapped in the pores of the salt.

The DTA diagram of $H_3PW_{12}O_{40}.21H_2O$ showed two endothermic peaks with maximum at 65 and 189 °C, and an exothermic one with maximum at 590 °C. The endothermic features may be ascribed to the release of physisorbed water and to the dehydration of $H_3PW_{12}O_{40}.6H_2O$ phase respectively, the same as the two weight losses below 250 °C observed in the TG diagram.

The exothermic peak may be assigned to the Keggin anion decomposition, which takes place without appreciable weight loss. Another weight loss, equivalent to 1.5 H_2O per Keggin unit, was observed in the TG diagram between 250–480 °C. It was assigned to evolution of constitutional water as a result of deprotonation of the heteropolyacid, as it fits well with the loss of the three protons of $H_3PW_{12}O_{40}$ [89]. During the water loss processes, Keggin anions are not too much disturbed and near 600 °C they are decomposed into the constituent oxides or, according to Mioc et al. [90], transformed into a new monophosphate bronze type compound PW_8O_{26}. The DTA diagrams of partially substituted $Cs(K)_XH_{3-X}PW_{12}O_{40}$ salts dried at 70 °C showed an endothermic peak with maximum at 50 °C and an exothermic one with maximum at 571 °C. The TG diagrams of these salts presented two weight losses, which occur below 200 °C (loss of physisorbed water) and between 250 and 480 °C (evolution of constitutional water).

The proton content evaluated from the weight loss in the temperature zone ascribed to the deprotonation step is presented in Table 13. This procedure has been already reported as an accurate measurement of heteropolycompound H^+ content. A good agreement with the values drawn from the synthesis stoichiometry was observed. Similar accordance was displayed when the H^+ amount of the partially substituted salts impregnated with TPA and dried at 70 °C was estimated by the weight loss (Table 13).

It must be noted that the DTA and TG diagrams of these TPA-impregnated salts showed similar features to those observed for the respective salts. However, the exothermic peak appeared at a slightly lower temperature (Table 14).

At first, one may think that the exothermic peak is a consequence of salt decomposition. As it was observed by FT-IR that the salts were stable at 620 °C and taking into account the assumption about the prepared samples may be considered as TPA dispersed on the neutral salts, we have ascribed the exothermic peak to decomposition of the acid present in the salts.

Table 13. Proton content of the partially substituted salts

Sample	H^+content per Keggin unit	H^+content per Keggin unit
	Expected[a]	Measured[b]
$Cs_{2.9}H_{0.1}PW_{12}O_{40}$	0.1	0.1
$K_{2.9}H_{0.1}PW_{12}O_{40}$	0.1	0.1
$Cs_{2.6}H_{0.4}PW_{12}O_{40}$	0.5	0.4
$K_{2.5}H_{0.5}PW_{12}O_{40}$	0.5	0.5

[a]Drawn from the synthesis stoichiometry.
[b]Evaluated by thermogravimetry.

Table 14. Proton content and TG exothermic peak temperature of the partially substituted salts impregnated with TPA

Sample	H^+content per Keggin unit	H^+content per Keggin unit	Exothermic peak (°C)
	Expected[a]	Measured[b]	
$Cs_{2.9}H_{0.1}PW_{12}O_{40}$	0.5	0.4	555
$K_{2.9}H_{0.1}PW_{12}O_{40}$	0.6	0.7	551
$Cs_{2.6}H_{0.4}PW_{12}O_{40}$	0.7	0.5	565
$K_{2.5}H_{0.5}PW_{12}O_{40}$	0.8	0.9	563

[a]Drawn from the synthesis stoichiometry.
[b]Evaluated by thermogravimetry.

Gayraud et al. [91] have also reported exothermic peaks near 588 °C for samples prepared by grinding mixtures of TPA and the cesium stoichiometric salt, at a lower temperature than the bulk acid (600 °C). They have ascribed them to crystallization of the constituent oxides after the destruction of the Keggin unit of the acid part of the samples.

The acidity measurements of the catalysts by means of potentiometric titration with n-butylamine let us to estimate the number of acid sites and their distribution [28]. Nevertheless, the meq amine/total meq H^+ or the area under the curve is an appropriate measure to carry out a comparison of the acidity of different samples. The titration curves obtained for the partially substituted salts $Cs(K)_{2.9}H_{0.1}PW_{12}$ presented very strong acid sites (Figure 7), Ei in the range 900–1000mV, displaying an acid strength higher than the less substituted $Cs_{2.6}H_{0.4}PW_{12}O_{40}$, and $K_{2.5}H_{0.5}PW_{12}O_{40}$ salts (Ei = 700–800 mV).

The titrated fraction of the total amount of acidic sites of cesium salts was higher than that of potassium salts, being higher for nearly stoichiometric salts. So, it decreased in the following order $Cs_{2.9}H_{0.1}PW_{12}O_{40} > K_{2.9}H_{0.1}PW_{12}O_{40} > Cs_{2.6}H_{0.4}PW_{12}O_{40} > K_{2.5}H_{0.5}PW_{12}O_{40}$.

In addition, the acid strength of all the samples was higher than that of the $H_3PW_{12}O_{40}.6H_2O$ (Ei = 620 mV). The higher acid strength of the partially substituted salts is related to a low hydration degree of the protons. At 70 °C, the protons are present as $H^+(H_2O)_2$ species in the $H_3PW_{12}O_{40}.6H_2O$ phase while the partially substituted salts might have reached a dehydrated state in which the protons may be as H_3O^+ species or bare protons [92]. The acid strength displayed by $H_3PW_{12}O_{40}.21H_2O$ (Ei = 538 mV) is lower than that of $H_3PW_{12}O_{40}.6H_2O$. This trend is in accordance with a report by Drago et al. [93] indicating that hydrated TPA is expected to be a weaker acid than the anhydrous acid due to the endothermic displacement of coordinated water. Thus, the stronger acidity displayed by the synthesized salts fits well with the presence of protons in a lower hydration state than that in bulk acid. In addition, the small differences observed between the acid strength of the different salts are related to the cation characteristics and the replacement degree.

The isopropanol dehydration proceeds quickly on acidic sites [94] and it has been used by several authors [94, 95] as a test reaction for comparing the acidity of different catalysts. Isopropanol decomposition can form either propene and water (dehydration) or acetone and hydrogen (dehydrogenation). At high temperatures (300–600 °C), rupture of the C–C bonds can occur to give alkanes, CO and CO_2 [96].

The results of the isopropanol decomposition catalyzed by the partially substituted salts together with the value obtained for bulk TPA are shown in Table 15.

Table 15. Turn over frequency of catalyzed isopropanol dehydration and acetylation of isoamyl alcohol with acetic acid

Catalyst	TOF_{IPA}	TOF_{ESTER}
TPA	1.23×10^{-2}	9.04×10^{-4}
$Cs_{2.9}H_{0.1}PW_{12}O_{40}$	1.62×10^{-1}	6.12×10^{-2}
$K_{2.9}H_{0.1}PW_{12}O_{40}$	9.87×10^{-2}	5.29×10^{-2}
$Cs_{2.6}H_{0.4}PW_{12}O_{40}$	4.42×10^{-2}	1.81×10^{-2}
$K_{2.5}H_{0.5}PW_{12}O_{40}$	3.21×10^{-2}	1.68×10^{-2}

TOF_{IPA}: (moles of transformed isopropanol)/(moles of H^+ in the catalyst) seg.
TOF_{ESTER}: (moles of transformed ester)/(moles of H^+ in the catalyst) seg.

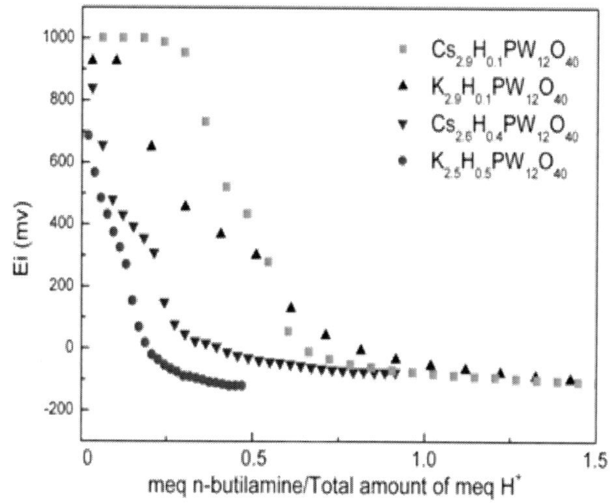

Figure 7. Potentiometric titration curves of the partially substituted cesium or potassium salts.

In the working conditions, we found that isopropanol decomposes to propene, only traces of acetone and diisopropylether were detected by means of GCMS. The specific dehydration conversion (TOF_{IPA}) of the different synthesized solids was higher than that of bulk TPA, due to the higher surface area of those catalysts allows a better access of the isopropanol molecules towards the protons.

Besides, the TOF_{IPA} of $Cs_{2.9}H_{0.1}PW_{12}O_{40}$ and $K_{2.9}H_{0.1}PW_{12}O_{40}$ salts were higher than that of $Cs_{2.6}H_{0.4}PW_{12}O_{40}$ and $K_{2.5}H_{0.5}PW_{12}O_{40}$, respectively. The order of conversion decrease ($Cs_{2.9}H_{0.1}PW_{12}O_{40} > K_{2.9}H_{0.1}PW_{12}O_{40} > Cs_{2.6}H_{0.4}PW_{12}O_{40} > K_{2.5}H_{0.5}PW_{12}O_{40}$) was similar to the above-mentioned results about acidic sites found from potentiometric titration and also to the decrease in the S_{BET} values.

The latter is in accordance with the fact that above 150 ºC the reactions catalyzed by supported HPA take place mainly on the surface or in the outer layers of the HPA crystals [97], which retain little amount of water, so they form a too rigid structure that isopropanol molecules cannot penetrate.

On the other hand, the TOF_{IPA} of samples obtained by incipient wetness impregnation of TPA on the partially substituted salts was lower than that of the parent compounds.

Thus, one may assume that protons of TPA added by impregnation do not enhance the activity for isopropanol dehydration at 180 ºC because they remain entrapped in the salt pores and are partially inaccessible to isopropanol molecules.

According to literature, esterification in vapor phase over various catalysts can be explained either through dual-site mechanism (Langmuir–Hinshelwood) or single-site mechanism (Rideal–Eley) [98-100]. Although the mechanism of homogeneously catalyzed liquid-phase esterification is long known [101], there is not still clearly explained the heterogeneously catalyzed gas-phase reaction. Chu et al. [102] found that the esterification mechanism of n-butanol with acetic acid catalyzed by immobilized dodecatungstosilicic acid on activated carbon proceeds via a protonated alcohol intermediate. However, a protonated carboxylic acid as the reaction intermediate has been proposed by other authors [103].

The catalytic activity (TOF_{ESTER}) of $Cs(K)_XH_{3-X}PW_{12}O_{40}$ salts for the acetylation of isoamyl alcohol with acetic acid was higher than that of bulk TPA (Table 15). This could be explained taking into account that the higher surface area lead to a higher amount of catalytically active protons. The catalysts showed selectivity near 80% to the ester. The remaining products detected by means of GCMS were 3-methyl-1-butene, 2-methyl-1-butene (as minor by-product) and traces of isovaleric aldehyde. The formation of 3-methyl-1-butene as the major product and 2-methyl-1-butene as minor product, together with the isovaleric aldehyde was found during the dehydration of isoamyl alcohol over a solid catalyst [103].

For the studied reaction, TOF_{ESTER} of the $Cs(K)_{2.9}H_{0.1}PW_{12}O_{40}$ salts is nearly thrice than that of the less substituted salts which may be related to enhanced proton dispersion. It can be observed that the specific conversion of the partially substituted salts decreased as S_{BET} also diminished. Moreover, a straight correlation of the catalytic behavior with the acidic properties of the catalysts was observed. Thus, TOF_{ESTER} presented a nearly linear relation with the conversion in the isopropanol dehydration test reaction both for the salts and the TPA-impregnated salts. This correlation together with the obtained by-products let us to assume that the acetylation occurs via a protonated alcohol intermediate, as proposed by Chu et al. [102]. The activity results shown here demonstrate that the synthesized catalysts, mainly the partially substituted salts, are interesting materials to obtain isoamyl acetate, an important product for a number of industries, as it is used for instance as flavor in mineral waters and syrups or perfume in diverse products.

In sum, the cesium or potassium salts obtained through partial neutralization of tungstophosphoric acid aqueous solutions may be considered as formed by well dispersed acid on the pore surface of the stoichiometric salt. The specific catalytic activity of the salts in the acetylation of isoamyl alcohol with acetic acid is straightly correlated with the acidic characteristics found through the isopropanol dehydration test reaction which, in turn, displayed the same order observed for the potentiometrically titrated acidic sites. The salts presented higher specific surface area than the bulk acid, thus leading to an increase in activity as the proton accessibility for the reactants is improved.

Conclusion

Acetic acid is one of the simplest carboxylic acids, and an important chemical reagent and industrial chemical for different uses.

Among the most common acetic acid uses it can be mentioned the manufacture of fragrances and perfumes, the use as a solvent for many industrial processes, the production of synthetic fibers and textiles, the manufacture of rubbers and plastics, in the production of pesticides, and it is also approved as a food additive.

We described here one specific application of its use, as acetylating agent, particularly in protection of alcohols and phenols. These reactions were performed in green conditions, using recyclable solid acids as catalysts. The protection of alcohols as the corresponding acetate esters is one of the most fundamental and widely used strategies in contemporary organic synthesis. Usually acetyl chloride and acetic anhydride are employed as the acylating agents in the presence of a convenient acidic or basic catalyst. However, both kinds of classical reagents are corrosive and lachrymator, and involve the use of tertiary amines such triethylamine, pyridine and DMAP, therefore they are not always ideal agents.

From the environmental point of view, a much convenient method for performing the acetylation reaction is the use of acetic acid as acetylating agent, because only water is produced as by-product.

Another advantage is the relatively cheaper route for industrial purposes when compared to the use of anhydrides or acid chlorides. The need for greener techniques leads to using different environmentally friendly reaction conditions; among them, the replacement of pollutant inorganic acid catalysts, such as sulfuric or hydrochloric acids, by reusable solid acids and the use of room temperature avoiding media heating are still necessary. The application of solid acids in organic transformation has an important role, because they have many advantages such as ease of handling, decreased plant corrosion, and more environmentally safe waste disposal procedures.

In this chapter the use of acetic acid as acetylating agent using recyclable solid acids synthesized in our laboratory as catalysts is reviewed. At first, alcohols and phenols acetylation using borated zirconia, obtained using different preparation conditions, as catalysts is presented. A very good agreement between the results obtained in the acetylation of 2-phenoxyethanol and the catalyst acidity was observed. The most appropriate catalyst was borated zirconia calcined at 320 °C containing 15 g B_2O_3/100 g of a support obtained using a high concentration of $ZrOCl_2$ precursor in the starting solution, achieving a yield of 80% in acetylated product at 14 h under reaction.

Besides, the catalysts kept their activity after three catalytic cycles. Then the acetylation of 2-phenoxyethanol with acetic acid employing borated zirconias modified with tungsten is presented, which led to good yields to acetylated product, unlike tungstated zirconias that gave low yields.

The better yields were reached with the modified catalysts obtained using borated zirconia prepared with a high concentration of zirconium precursor as support.

For both types of catalysts, the yield to acetylated product obtained for a number of different alcohols and phenols is examined. The alcohol and phenol structure is a key in the yield to the acetylation product. The reactivity order toward acetylation with acetic acid using these catalysts depended on two factors, an inductive effect and a steric effect.

Finally, the isoamyl alcohol acetylation using tungstophosphoric acid supported on zirconia; and partially substituted Cs and K salts of tungstophosphoric acid is examined. The catalyst obtained by incipient wetness impregnation of zirconia with tungstophosphoric is the most active catalyst, for the acetylation reaction which leads to isoamyl acetate, thus obtaining an important product for flavor and fragrance industries.

Similarly, the cesium or potassium salts obtained through partial neutralization of tungstophosphoric acid aqueous solution showed excellent activity and was straightly correlated with the acidic characteristics found through the isopropanol dehydration test reaction.

In sum, it may be concluded that very simple and convenient catalytic methods for the acetylation of alcohols and phenol, using acetic acid as acetylation agent, were developed. They have the advantages of giving very good yields and excellent selectivity, at reasonable reaction time, thus leading to low environmental impact, both due to the recyclable solid acids and the clean acetylation agent.

Appendix

Chemicals were purchased from Aldrich, Fluka and Merck chemical companies and were freshly used after purification by standard procedures (distillation and recrystallization). All the reactions were monitored by TLC on precoated silica gel plates (254 mm). Flash column chromatography was performed with 230 to 400 mesh silica gel. All the yields were calculated from crystallized products. All the products were identified by comparison physical data (TLC and NMR) with those reported or with those of authentic samples prepared by the respective conventional methods using sulfuric acid as catalyst. The ^{1}H-NMR and ^{13}C-NMR spectra were obtained by a Bruker instrument 400 MHz model as $CDCl_3$ solutions, and the chemical shifts were expressed in d units with Me4Si (TMS) as the internal standard.

For the preparation of borated zirconias, firstly the supports were obtained from aqueous solutions of $ZrOCl_2.8H_2O$ with a concentration of 0.06 and 0.56 mmoles Zr/cm^3, to which a 30% NH_4OH solution was added dropwise until a final pH of 10. The obtained products were aged for 72 h and filtered in vacuum. They were then washed with distilled water up to chloride elimination and dried at room temperature, giving the S2 and S4 materials, respectively. These solids were impregnated with aqueous solutions of H_3BO_3 with 0.05, 0.3 or 0.7 mmoles B/cm^3 concentration, in order to obtain boron concentrations in the solid of 2.5, 15 or 35, expressed as g B_2O_3/100 g support. Then, the suspension was evaporated till dryness on a flat plate at low temperature. Afterward, the solids were calcined under N_2 atmosphere at 320 °C for 5 h. The samples thus obtained will be named SXBY, where X is the number identifying the support and Y is the boron concentration. A portion of the S4B35 sample was also calcined at 500 and 750 °C.

To obtain the borated zirconias modified with ammonium metatungstate, the supports named S2 and S4 were impregnated with an aqueous solution containing 0.3 moles B/dm^3, prepared from H_3BO_3, for 12 h, under stirring, in order to obtain a boron concentration in the solid of 15 % (w/w), expressed as g B_2O_3/100 g support. Afterwards, the solution was evaporated and the solids were dried in an oven and, finally, were calcined at 320 °C for 5 h under N_2 atmosphere (S2B15 and S4B15 samples).

Each sample of borated zirconia previously obtained was divided into two portions, which were impregnated with an aqueous solution containing 0.11 or 0.22 moles W/dm^3, prepared from ammonium metatungstate (AMT), in order to obtain a tungsten concentration in the solid of 12.5 % or 25 % (w/w), expressed as g W/100 g support. The impregnation was carried out under stirring for 12 h. After evaporation of the excess solution, the solids were dried in an oven, and calcined at 320 °C for 5 h under N_2 atmosphere. The samples S2B15W12.5, S2B15W25, S4B15W12.5 and S4B15W25 were thus obtained. At the same time, solids were prepared with the same tungsten contents in zirconia, without boron addition, under the same conditions as the previous samples, thus obtaining the materials S2W12.5, S2W25, S4W12.5 and S4W25.

For the synthesis of the catalysts based on tungstophosphoric acid supported on zirconia, zirconium hydroxide gel was prepared from 0.07 M zirconium oxychloride aqueous solution by adding dropwise an ammonium hydroxide solution (25% ammonia) up to pH = 9.5. The hydrogel was aged at room temperature (r.t.) for 144 h, then filtered and washed with deionized water and finally calcined for 24 h at 310 °C (named Z hereinafter). The support was impregnated by means of the incipient wetness method with ethanol–water (e-w) solution

(50 %, v/v) of TPA Fluka ($H_3PW_{12}O_{40}\cdot nH_2O$) with 120 g-W/dm^3 concentration. Then, after the filling of the pores, the sample was dried at 70 °C for 24 h, and named as Z-TPA-W. The support was also impregnated by an equilibrium adsorption method at 20 °C by contacting 1 g of Z with 4 ml of a 120 g-W/dm^3 TPA solution in e-w, under constant stirring during 72 h. This time was long enough to attain equilibrium of the adsorption–desorption processes. The solid was separated from the solution by centrifugation and then dried at 70 °C for 24 h, and named as Z-TPA-E.

To obtain the materials based on $H_3PW_{12}O_{40}$ supported on their partially substituted Cs or K salts, firstly the cesium or potassium salts were synthesized by adding slowly, under vigorous stirring, 5M solution of CsCl (Merck p.a.) or KCl (Merck p.a.) into a 0.1M aqueous solution of $H_3PW_{12}O_{40}$ (Fluka p.a.) in the desired stoichiometric proportion. The precipitates were separated from the solutions by centrifugation and were washed up to complete elimination of chlorides. They were then dried at 70 °C for 24 h. The salts obtained by this procedure will be designated herein after as $Cs(K)_XH_{3-X}PW_{12}O_{40}$, where X is the amount of protons per $[PW_{12}O_{40}]^{3-}$ anion in the salt. Afterwards, the salts were impregnated, using the incipient wetness method, with a $H_3PW_{12}O_{40}$ solution in ethanol–water (50%, v/v), in order to add 4.54 10^{-5} mol of $H_3PW_{12}O_{40}$ per gram of salt. The supported catalysts were dried at 70 °C for 24 h. The catalysts characterization was done by different physical-chemical techniques. The specific surface area of the samples was estimated by the BET method from the nitrogen adsorption–desorption isotherms at −196 °C, performed using Micromeritics equipment, model ASAP 2020, previously degassed at 100 °C for 1 h.

To obtain the spectra by Fourier transform infrared spectroscopy (FT-IR), Bruker IFS 66 equipment and pellets of the sample in KBr were used. The measurements were performed in the 400–4000 cm^{-1} range.

The thermogravimetric and differential thermal analysis (TG-DTA) was carried out using Shimadzu DT 50 thermal analyzer. The TG–DTA experiments were performed under argon or nitrogen respectively, using 25–50 mg samples and a heating rate of 10 °C/min. Quartz cells were used as sample holders with α-Al2O3 as reference, and the temperature range was 25–700 °C.

The spectra obtained by diffuse reflectance spectroscopy (DRS) were recorded in the range 200-600 nm, using a UV-visible Varian Super Scan 3 equipment, fitted with a diffuse reflectance chamber with inner surface of $BaSO_4$. Samples were compacted in a teflon sample holder to obtain a sample thickness of 2 mm.

The X-ray diffraction (XRD) patterns were obtained with Philips PW-1417 equipment, using Cu Kα radiation and Ni filter, 20 mA and 40 kV in the high tension source and a scanning rate of 1° per min.

For the characterization of the solids by ^{31}P nuclear magnetic resonance spectroscopy (^{31}P MAS-NMR), a Bruker MSL-300 equipment was employed, using 5 μs pulses, a repetition time of 3 s and working at a frequency of 121.496 MHz for 31P at r.t., the resolution being 3.052 Hz per point. Phosphoric acid 85% was the external reference. A 5 mm diameter and 10 mm in height sample holder was used, the spin rate was 2.1 kHz. Several hundred pulse responses were collected.

The acidity determination by isopropanol dehydration was carried out in a conventional flow fixed bed reactor at atmospheric pressure and 180 ◦C. The gas carrier was helium and the isopropanol volumetric flow was 2.1 10^{-3} cm^3/s in a total flow of 50 cm^3/min.The reaction products were quantified by gas chromatography, using a thermal conductivity detector. The specific isopropanol conversion (as turnover frequency) was calculated under differential conversion conditions, that is conversions lower than 10%. It was expressed as moles of transformed isopropyl alcohol/(moles of H^+ in the solid) sec (TOF_{IPA}).

For the acidity determination by potentiometric titration, 0.05 g of solid was suspended in 45 cm^3 of acetonitrile. The suspension was stirred for 3 h, and then titrated with a 0.05 N solution of n-butylamine in acetonitrile at 0.05 cm^3/min. The electrode potential variation was measured with an Instrumentalia S.R.L. digital pHmeter, using a double junction electrode. The acidic properties of the samples measured by this technique enable the evaluation of the total number of acid sites and their acid strength. In order to interpret the results, it is suggested that the initial electrode potential (Ei) indicates the maximum acid strength of the surface sites, and the values (meq. amine/g solid) where the plateau is reached, indicates the total number of acid sites.

For the tests of acetylation of alcohols and phenols using borated zirconias or borated zirconias modified with ammonium metatungstate as catalysts, the reaction was carried out in a batch glass reactor equipped with a reflux condenser. A mixture of the corresponding alcohol or phenol (1 mmol), acetic acid (5 mmol), toluene (5 cm^3) and the catalyst (100 mg) was magnetically stirred at 110 °C up to a final reaction time of 14 h. Then, the catalyst was separated by filtration and washed twice with a small quantity of toluene (1 cm^3). The progress of the reaction was monitored by silica gel thin layer chromatography. The organic solution was washed twice with 5% $NaHCO_3$ solution (3 cm^3), with water (3 cm^3), and dried over anhydrous Na_2SO_4. The solution was then concentrated under vacuum, and the residue was purified by flash column chromatography using silica gel 230–400 mesh as the stationary phase and a mixture of n-hexane and ethyl acetate of growing polarity as the elution solvent. The isolated products were characterized by spectral studies (^1H and ^{13}C NMR). The spectra were compared with those of standard esters. The yield was calculated as mmol of purified product with respect to mmol of initial alcohol (phenol). For the experiments of reuse, the catalyst was treated after reaction by washing twice with hot toluene (3 cm^3), dried under vacuum at 20 °C for 5 h and then reused.

The test reaction of isoamyl alcohol acetylation using TPA supported on ZrO_2 or $H_3PW_{12}O_{40}$ partially substituted Cs or K salts as catalysts was carried out in a conventional flow fixed-bed reactor, operating at atmospheric pressure, with a solid mass of 0.5 g. The gas employed as carrier was He. The acetic acid–isoamyl alcohol mixture (1:1 molar ratio) volumetric flow was 0.13 cm^3/min in a total flow of 50 cm^3/min.

The temperature was set at 180 °C and both the reactants and reaction products were quantified by gas chromatography, using a thermal conductivity detector. The turnover number (TON) was calculated as the ratio between moles of ester formed at 4 h of reaction and moles of HPA in the catalysts. The turnover frequency (TOF_{ESTER}), calculated under differential conversion conditions, was expressed as moles of formed ester/(moles of H^+ in the catalyst) sec.

Acknowledgments

The authors thank L. Osiglio and A. G. Sathicq for their experimental contribution and to CONICET, UNLP and ANPCyT for the financial support.

References

[1] Z. Xu, Z. Shi, L. Jiang, In: *Comprehensive Biotechnology* (Second Edition), Vol. 3: Industrial Biotechnology and Commodity Products, Elsevier B.V., Editor-in-Chief: Murray Moo-Young, 2011, 189–199.

[2] G. Sartori, R. Ballini, F. Bigi, G. Bosica, R. Maggi, P. Righi, *Chem. Rev.* 104 (2004) 199-250.

[3] T. Reddy, M. Narasimhulu, N. Suryakiran, Ch. Mahesh, K. Ashalatha, Y. Venkateswarlu, *Tetrahedron Lett.* 47 (2006) 6825-6829.

[4] M. Heravi, F. Behbahani, V. Zadsirjan, H. Oskooie, *J. Braz. Chem. Soc.* 17 (2006) 1045-1047.

[5] L. Osiglio, G. Romanelli, M. Blanco, *J. Mol. Catal. A: Chem.* 316 (2010) 52-58.

[6] R. Zhdanov, S. Zhenodarova, *Synthesis* 4 (1975) 222-245.

[7] E. Scriven, *Chem. Soc. Rev.* 12 (1983) 129-133.

[8] G. Hofle, V. Steglich, H. Vorbruggen, *Angew. Chem. Int. Ed. Engl.* 17 (1978) 569-583.

[9] A. Cope, E. Herrick, *Org. Synth.* 4 (1963) 304.

[10] R. Baker, F. Bordwell, *Org. Synth.* 3 (1955) 141.

[11] S. Chandrasekhar, T. Ramachander, M. Takhi, *Tetrahedron Lett.* 39 (1998) 3263-3266.

[12] G. Bartoli, M. Bosco, R. Dalpozzo, E. Marcantoni, M. Massaccesi, L. Sambri, *Eur. J. Org. Chem.* 23 (2003) 4611-4617.

[13] K. Ishihara, M. Kubota, H. Kurihara, H. Yamamoto, *J. Am. Chem. Soc.* 117 (1995) 4413-4414.

[14] K. Chauhan, C. Frost, I. Love, D. Waite, *Synlett* 11 (1999) 1743-1744.

[15] P. Saravanan, V. Singh, *Tetrahedron Lett.* 40 (1999) 2611-2614.

[16] S. De, *Tetrahedron Lett.* 45 (2004) 2919-2922.

[17] A. Rahmatpour, *Comp. Rend. Chimie*, 15 (2012) 1048-1054.

[18] N. Salavati-Niasari, *Comb. Chem. High Through. Screen.* 15 (2012) 705-712.

[19] D. Zareyee, A. Ghadikolaee, M. Khalilzadeh, *Can. J. Chem.* 90 (2012) 464-468.

[20] G. Romanelli, D. Bennardi, J. Autino, G. Baronetti, H. Thomas, *E-J. Chem.* 5 (2008) 641-647.

[21] P. Patil, K. Malshe, P. Kumar, M. Dongare, E. Kemnitz, *Catal. Commun.* 3 (2002) 411-416.

[22] B. Xu, S. Cheng, S. Jiang, Q. Zhu, *Appl. Catal. A: Gen.* 188 (1999) 361-368.

[23] B. Madje, P. Patil, S. Shindalkar, S. Benjamin, M. Shingare, M. Dongare, *Catal. Commun.* 5 (2004) 353-357.

[24] K. Malshe, P. Patil, S. Umbarkar, M. Dongare, *J. Mol. Catal. A: Chem.* 212 (2004) 337-344.

[25] T. Yamaguchi, *Catal. Today* 20 (1994) 199-218.

[26] B.-Y. Zhao, X.-P. Xu, H.-R. Ma, D.-H. Sun, J.-M. Gao, *Catal. Lett.* 45 (1997) 237-244.

[27] J. C. Duchet, M. J. Tilliette, D. Cornet, *Catal. Today* 10 (1991) 507-520.

[28] R. Cid, G. Pecci, *Appl. Catal.* 14 (1985) 15-21.

[29] D. B. Ravindra, Y. T. Nie, S. Jaenicke, G. K. Chuah, *Catal. Today* 96 (2004) 147-153.

[30] K. Hadjiivanov, J.-C. Lavalley, *Catal. Commun.* 2 (2001) 129-133.

[31] A. Khder, E. El-Sharkawy, S. El-Hakam, A. Ahmed, *Catal. Commun.* 9 (2008) 769-777.

[32] S. Barbosa, M. Dabdoub, G. Hurtado, S. Klein, A. Baroni, C. Cunha, *Appl. Catal. A: Gen.* 313 (2006) 146-150.
[33] Y. Wang, W. Li, *React. Kinet. Catal. Lett.* 83 (2004) 195-203.
[34] M. B. Smith, J. March, *Advanced Organic Chemistry. Reactions, Mechanisms and Structure,* 6th ed., Wiley Interscience Publication, US, 2007.
[35] V. Mirkhani, S. Tangestaninejad, M. Moghadam, B. Yadollahi, L. Alipanah, *Monatsh. Chem.* 135 (2004) 1257-1263.
[36] M. A. Cortés-Jácome, C. Angeles-Chavez, E. López-Salinas, J. Navarrete, P. Toribio, J. A. Toledo, *Appl. Catal. A: Gen.* 318 (2007) 178-189.
[37] J. D. Hansen, A. L. Kustov, S. B. Rasmussen, R. Fehrmann, C. H. Christesen, *Appl. Catal. B: Environ.* 66 (2006) 161-167.
[38] P. Wongmaneenil, B. Jongsomjit, P. Praserthdam, *Catal. Commun.* 10 (2009) 1079-1084.
[39] K. Ngaosuwan, X. Mo, J. G. Goodwin Jr., P. Praserthdam, *Appl. Catal. A: Gen.* 380 (2010) 81-86.
[40] C. D. Baertsch, K. T. Komala, Y.-H. Chua, E. Iglesia, *J. Catal.* 205 (2002) 44-57.
[41] R. A. Boyse, E. Ko, *J. Catal.* 171 (1997) 191-207.
[42] B. M. Reddy, M. K. Patil, G. K. Reddy, B. T. Reddy, K. N. Rao, *Appl. Catal. A: Gen.* 332 (2007) 183-191.
[43] S. Ramu, N. Lingaiah, B. L. A. Prabhavathi, R. B. N. Prasad, I. Suryanarayana, P. S. Sai Prasad, *Appl. Catal. A: Gen.* 276 (2004) 163-168.
[44] E. I. Ross-Medgaarden, W. V. Knowles, T. Kim, M. S. Wong, W. Zhou, C. J. Kiely, I. E. Wachs, *J. Catal.* 256 (2008) 108-125.
[45] L. Osiglio, A. Sathicq, G. Romanelli, M. Blanco, *J. Molec. Catal. A: Chem.* 359 (2012) 97-103.
[46] K. Arata, H. Nakamura, M. Shouji, *Appl. Catal. A: Gen.* 197 (2000) 213-219.
[47] R. Sakthivel, H. Prescott, E. Kemnitz, *J. Mol. Catal. A: Chem.* 223 (2004) 137-142.
[48] A. Bordoloi, N. T. Mathew, B. M. Devassy, S. P. Mirajkar, S. B. Halligudi, *J. Mol. Catal. A: Chem.* 247 (2006) 58-64.
[49] T. Green, P. Wutz, *Protective Groups in Organic Synthesis*, Wiley, New York, 1991.
[50] M. Alizadeh, T. Kermani, R. Tayebee, *Monatsh fur Chemie* 138 (2007) 165-170.
[51] J. F. Keggin, *Proc. Roy. Soc. London A* 144 (1934) 75-100.
[52] T. Okuhara, N. Mizuno and M. Misono, In: *Advances in Catalysis*, Vol. 41, eds. D. D. Eley, W. O. Haag and B. Gates, Academic Press, New York, 1996.
[53] Y. Izumi, K. Urabe and M. Onaka, Zeolite, Clay and Heteropoly Acid in Organic Reactions, Kodansha/VCH, Tokyo, 1992.
[54] I. Kozhevnikov, *Catal. Rev. Sci. Eng.* 37 (1995) 311-352.
[55] P. Vázquez, M. Blanco, C. Cáceres, *Catal. Lett.* 60 (1999) 205-215.
[56] L. Pizzio, C. Cáceres, M. Blanco, *Appl. Catal.* A 167 (1998) 283-294, and references therein.
[57] E. López-Salinas, J. Hernández-Cortez, I. Schiffer, E. Torres García, J. Navarrete, A. Gutiérrez-Carrillo, T. López, P. P. Lottici, D. Bersani, *Appl. Catal. A: Gen.* 193 (2000) 215-225.
[58] Y. Izumi, K. Urabe, *Chem. Lett.* (1981) 663-666.
[59] S. Kasztelan, E. Payen, J. B. Moffat, *J. Catal.* 112 (1988) 320-324.
[60] D. A. Ward, E. I. Ko, *J. Catal.* 150 (1994) 18-33.

[61] L. M. Kustov, V. B. Kazansky, F. Figueras, D. Tichit, *J. Catal.* 150 (1994) 143-149.
[62] R. A. Boyse, E. I. Ko, *Catal. Lett.* 38 (1996) 225-230.
[63] L. Pizzio, P. Vázquez, C. Cáceres, M. Blanco, *Stud. Surf. Sci. Catal.* 130 (2000) 953-958.
[64] L. Pizzio, P. Vázquez, C. Cáceres, M. Blanco, *Catal. Lett.*77 (4) (2001) 233-239.
[65] M. Fournier, R. Thouvenot, C. Rocchiccioli-Deltcheff, *J. Chem. Soc. Faraday Trans.* 87 (1991) 349-356.
[66] K. Nomiya, Y. Sugi, K. Amimoto, M. Miwa, *Polyhedron* 6 (1987) 519-524.
[67] P. Vázquez, M. González, M. Blanco, C. Cáceres, *Stud. Surf. Sci. Catal.* 91 (1995) 1121-1130.
[68] C. Rocchiccioli-Deltcheff, M. Fournier, R. Franck, R. Thouvenot, *Inorg. Chem.* 22 (1983) 207-216.
[69] D. C. Vermaire, P. C. van Berge, *J. Catal.* 116 (1989) 309-317.
[70] S. M. Ahmed, *J. Phys. Chem.* 73 (1969) 3546-3555.
[71] M. P. McDaniel, *J. Phys. Chem.* 77 (1973) 2070.
[72] G. A. Parks, *Chem. Rev.* 65 (1965) 177-198.
[73] C. Vera, J. Parera, *J. Catal.* 165 (1997) 254-262.
[74] C. Rocchiccioli-Deltcheff, R. Thouvenot, R. Franck, *Spectrochim. Acta* 32A (1976) 587-597.
[75] C. Rocchiccioli-Deltcheff, R. Thouvenot, *Spectr. Lett.* 12 (1979) 127-138.
[76] R. Massart, R. Contant, J. M. Fruchart, J. P. Ciabrini, M. Fournier, *Inorg. Chem.* 16 (1977) 2916-2921.
[77] J. B. McMonagle, J. B. Moffat, *J. Colloid Interface Sci.* 101 (1984) 479-488.
[78] E. M. Serwicka, K. Bruckman, J. Haber, E. A. Paukshtis, E. N. Yurchenko, *Appl. Catal.* 73 (1991) 153-163.
[79] L. R. Pizzio, C. V. Cáceres, M. N. Blanco, *Stud. Surf. Sci. Catal.* 143 (2002) 731-738.
[80] L. Pizzio, M. Blanco, *Appl. Catal. A: Gen.* 255 (2003) 265-277.
[81] N. Essayem, G. Coudurier, M. Fournier, J. Védrine, *Catal. Lett.* 34 (1995) 223-235.
[82] P. Y. Gayraud, I. H. Stewart, S. B. Derouane-Abd Hamid, N. Essayem, E. G. Derouane, J. C. Védrine, *Catal. Today* 63 (2000) 223-228.
[83] N. Essayem, Y. Ben Taârit, P. Y. Gayraud, G. Sapaly, C. Naccache, *J. Catal.* 204 (2001) 157-162.
[84] M. A. Parent, J. B. Moffat, *Catal. Lett.* 48 (1997) 135-143.
[85] S. Choi, Y. Wang, Z. Nie, J. Liu, C. H. F. Peden, *Catal. Today* 55 (2000) 117-124.
[86] G. M. Brown, M. R. Noe-Spirlet, W. R. Busing, H. A. Levy, *Acta Crystallogr.* B 33 (1977) 1038-1046.
[87] M. A. Parent, J. B. Moffat, *Langmuir* 12 (1996) 3733-3739.
[88] Y. Kanda, K. Y. Lee, S. Nakata, S. Asaoka, M. Misono, *Chem. Lett.* (1989) 139.
[89] M. Fournier, C. Feumi-Jantou, C. Rabia, G. Hervé, S. Launary, *J. Mater. Chem.* 2 (1992) 971-978.
[90] J. B. Mioc, R. Z. Dimitrijevi, M. Davidovic, Z. P. Nedic, M. M. Mitrovic, P. H. Colomban, *J. Mater. Sci.* 29 (1994) 3705-3718.
[91] P. Y. Gayraud, N. Essayem, J. C. Védrine, *Catal. Lett.* 56 (1998) 35-41.
[92] N. Essayem, Y. Y. Tong, H. Jobic, J. C. Vedrine, *Appl. Catal. A: Gen.* 194/5 (2000) 109-122.
[93] R. S. Drago, J. A. Dias, T. O. Maier, *J. Am. Chem. Soc.* 119 (1997) 7702-7707.

[94] J. R. Sohn, H. J. Jang, *J. Mol. Catal.* 64 (1991) 349-360.
[95] M. Ai, S. Susuki, *J. Catal.* 30 (1973) 362-371.
[96] O. V. Krylov, *Catalysis by Nonmetals*, Academic Press, New York, 1970, p. 115.
[97] Y. Izumi, R. Hasebe, K. Urabe, *J. Catal.* 84 (1983) 402-404.
[98] J. C. Kuriacose, S. S. Jewur, *Int. J. Chem. Kinet.* 9 (1977) 641-650.
[99] J. Giménez, J. Costa, S. Cervera-March, *Appl. Catal.* 31 (1987) 221-234.
[100] E. Santacesaria, D. Gelosa, P. Danise, S. Carra, *J. Catal.* 80 (1983) 427-436.
[101] I. Roberts, H. C. Urey, *J. Am. Chem. Soc.* 60 (1938) 2391-2393.
[102] W. Chu, X. Yang, X. Ye, Y. Wu, *Appl. Catal. A* 145 (1996) 125-140.
[103] G. P. Babu, R. S. Murthy, V. Krishnan, *J. Catal.* 166 (1997) 111-114.

In: Acetic Acids
Editor: Angelo Basile

ISBN: 978-1-62948-217-0

Chapter 5

Hydrogen Production via Acetic Acid Steam Reforming Reaction Using Membrane Reactors

A. Iulianelli, S. Liguori, F. Dalena, and A. Basile*
CNR-ITM, c/o University of Calabria, Rende (CS), Italy

Abstract

In recent years, hydrogen production from biomass transformation has been very attractive. Particularly interesting are methods in which the biomass is converted into intermediate liquid bio-fuels such as pyrolysis oil, acetic acid or ethanol. Among various renewable sources, acetic acid may be easily produced by fermentation of biomass. Furthermore, acetic acid, unlike methanol and ethanol, is not inflammable and, as a consequence, it represents a safe hydrogen carrier.

In particular, acetic acid may be easily transformed into hydrogen with high selectivity over effective catalysts such as Ni–Co and Ru. However, to the best of our knowledge a restricted literature is noticeable on hydrogen production from acetic acid and only a few manuscripts deal with conventional reactors usually operating at temperature higher than 500 °C. In these studies, acetic acid steam reforming reaction is carried out using different catalysts and a hydrogen rich-gas stream is produced, containing secondary products such as CO_2, CO, CH_4, etc. Therefore, with the purpose of producing CO_x-free hydrogen for PEM fuel cell applications, the hydrogen rich-gas stream needs to be purified. This chapter is aimed to study the utilization of inorganic membrane reactor technology for producing CO_x-free hydrogen in only one step from acetic acid steam reforming instead of the conventional reactors followed by further stages for hydrogen purification such as water gas shift reactors (high and low temperature), partial oxidation reactors and pressure swing adsorption devices.

* Corresponding author: A. Basile. CNR-ITM, c/o University of Calabria, Via Pietro Bucci, Cubo 17/C, 87036 Rende (CS), Italy. E-mail: a.basile@itm.cnr.it; Phone: +39 0984 492013; Fax: +39 0984 402103.

Abbreviations

FBR	Fixed bed reactor
FluBR	Fluidized bed reactor
CR	Catalytic reactor
MR	Membrane reactor
HT	High temperature
LT	Low temperature
PROX	Preferential oxidation
WGS	Water gas shift
PSA	Pressure Swing Adsorption
AASR	Acetic acid steam reforming
AcAc	Acetic acid
PEMFC	Polymer electrolyte membrane fuel cell
J_{H2}	Hydrogen flux permeating through the membrane
Pe	Hydrogen permeability
δ	Membrane thickness
$p_{H2\text{-retentate}}$	Hydrogen partial pressures in the retentate side
$p_{H2\text{-permeate}}$	Hydrogen partial pressures in the permeate side
n	Dependence factor of the hydrogen flux on the hydrogen partial pressure
GHSV	Gas hourly space velocity
WHSV	Weight hourly space velocity
LHSV	Liquid hourly space velocity

Introduction

In the last decade, it has become imperative to limit the exploitation of derived of fossil fuels owing to the emission of CO_2 that represents one of the main causes of global warming, also inducing a severe climate change [1]. With the aim of reducing the derivation of fossil fuels dependency and, consequently, CO_2 emissions, many efforts from Academia and Industry are devoted to investigate new and/or alternative fuels and sustainable energy sources. Particular interest has received the exploitation of biomass, which consists of a renewable and CO_2-neutral fuel carbon source [2]. Fischer and Schrattenholzer [3] estimated the potentiality of biomass energy exploitation in 2050, which should cover an energy need up to 1100 EJ/year versus an estimated total global energy consumption between 600 and 1000 EJ/year, completely matching the energy request of the world with the advantage of avoiding the derived fossil fuels exploitation.

It is well known that biomass can be converted by thermal and catalytic processes. For example, it can be gasified to produce synthesis gas, which can be further utilized in gas turbines for power production or purified for reaching a high grade hydrogen [4].

Another interesting transformation of biomass is the flash pyrolysis, which makes possible the production of bio-oil, besides gases and solids (char) [5].

In particular, bio-oil shows a volumetric energy density up to ten times larger than biomass, making advantageous its transportation as well as more cost effective with respect to the pristine biomass [6]. Bio-oil compounds distribution is constituted by several oxygenated compounds like acids, ketones, alcohols, phenols, etc. This distribution is variable and, generally, depends on which kind of feedstock is used and on the conditions of bio-oil production. In detail, acids and lignin represent the largest organic fraction, even if a significant presence of water is noticeable.

However, acetic acid, hydroxyacetaldehyde, acetol and levoglusacon constitute some of the most abundant molecules in bio-oil [7]. One of the possible drawbacks of bio-oil utilization in catalytic processes is represented by the sulfur content, particularly when Ni or noble metals-based catalysts are used.

Among the various bio-oil compounds, acetic acid is renewable and can be easily obtained by fermentation. In addition, acetic acid, unlike methanol and ethanol, is not inflammable, then representing a safe hydrogen carrier.

Indeed, in this chapter, special attention is devoted to acetic acid exploitation (as a product coming from bio-oil) in steam reforming process for hydrogen production. An overview on the conventional systems to carry out the reforming reaction and about the most used catalysts is also given.

Furthermore, membrane reactor (MR) technology is introduced to offer a view on the alternative solutions to the conventional reactors (CRs) for hydrogen production from acetic acid steam reforming (AASR) reaction and for pointing out the main benefits of this technology as well as the future perspectives at larger scale.

1. Steam Reforming of Acetic Acid

As mentioned above, the presence of carboxylic acids is large in the bio-oil and acetic acid represents the most abundant fraction [8]. As a consequence, acetic acid has been used as a model compound of bio-oil in many studies and there are several studies dealing with AASR reaction (1) [9-27].

$$CH_3COOH + 2H_2O = 2CO_2 + 4H_2 \quad (1)$$

Unfortunately, AASR reaction includes also undesired reactions as reported below:

$$2CH_3COOH = CH_3COCH_3 + CO_2 + H_2O \text{ (ketonization)} \quad (2)$$

$$CH_3COOH = CH_2CO + H_2O \text{ (dehydration)} \quad (3)$$

The aforementioned secondary reactions are the ketonization (2) to acetone and the dehydration to ketene (3). In particular, acetone can react via polymerization via aldol condensation, making possible coke formation from reactions (4-6) [23]; while ketene can react to form ethene by ketene coupling reaction (7), which is a coke precursor [28]:

$$2CH_3COCH_3 \rightarrow H_2O + (CH_3)_2CCHCOCH_3 \quad (4)$$

$$(CH_3)_2CCHCOCH_3 + CH_3COCH_3 \rightarrow C_9H_{12} + 2H_2O \quad (5)$$

$$C_9H_{12} \rightarrow \text{Coke} \quad (6)$$

$$2CH_2CO = C_2H_4 + 2CO \text{ (ketene coupling)} \quad (7)$$

However, both acetone and ketene, derived respectively from reactions (2) and (3), can be converted in H_2 and CO via steam reforming reactions as represented by (8) and (9), wchich are considered desired reactions.

$$CH_3COCH_3 + 3H_2O = 3CO + 6H_2 \text{ (acetone steam reforming)} \quad (8)$$

$$CH_2CO + H_2O = 2CO + 2H_2 \text{ (ketene steam reforming)} \quad (9)$$

Other desired reactions are the decomposition of acetic acid in CH_4 (10) and CO_2 (11):

$$CH_3COOH = CH_4 + CO_2 \quad (10)$$

$$2CH_3COOH = 2H_2 + CO_2 + C \quad (11)$$

Various authors [10, 23, 28, 29] pointed out that complete acetic acid conversion can be reached in the range from 400 to 800 °C, depending on the catalyst and operating conditions utilized. Basagiannis et al. [28] demonstrated that, by using catalysts without a catalytic active metal such as Al_2O_3, acetic acid conversion exceeds 10% at T > 850 °C and it is mainly due to thermal decomposition reactions to CH_4 and CO_2. In the meanwhile, at higher temperatures (around 900 °C or more), hydrogen production can also take place over Al_2O_3.

However, acetic acid conversion over quartz particles is higher than that achieved by using support materials and, more in detail, complete conversion can be reached at 950 °C [10, 28].

1.1. Catalysts for AASR Reaction to Produce Hydrogen

Hydrogen can be produced easily from AASR reaction with high selectivity by using effective catalysts such as Pt, Ru, Rh, Ni and Co [5, 30, 31]. However, there is not a comprehensive literature on the argument and most of the scientific publications deal with conventional fixed bed reactors (FBRs), Figure 1.

In particular, Hu et al. [10, 11, 13, 14] studied AASR in FBRs to produce hydrogen paying attention to the utilization of Ni, Co, Fe and Cu as metal phases supported on Al_2O_3. Furthermore, they investigated also the effects due to the variation of the support in Ni-based catalysts [12, 15].

As a main result, Hu and co-workers reached 100% acetic acid conversion by using Ni-Co based catalysts, also obtaining hydrogen as a principal product [12]. Nevertheless, these authors demonstrated that Ni/Al_2O_3 and Co/Al_2O_3 catalysts are active for AASR, while Fe/Al_2O_3 and Cu/Al_2O_3 catalysts present negligible activity [10].

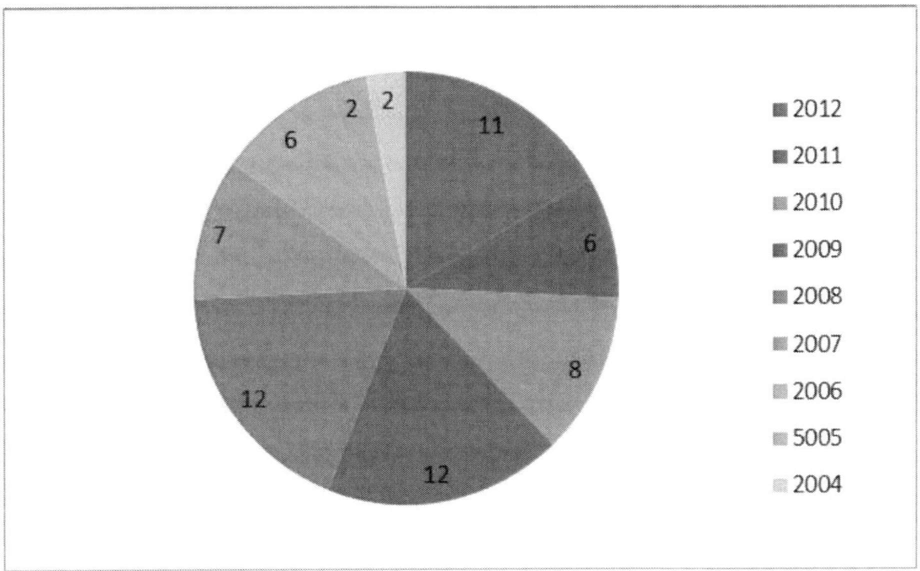

Figure 1. Number of publications on acetic acid steam reforming in FBRs vs year. Scopus database: www.scopus.com.

Li et al. [18, 19] used Ni/ZrO_2 for carrying out AASR achieving ~ 90 of conversion at 500 °C and 16wt% of metal loading. The authors pointed out that, by using Ni/ZrO_2 catalyst, the polymerization of acetone is the main routes for coke deposition even though coke formation is suppressed with the increase of nickel loading up to 16 wt.%, while coke formation restarts to increase for higher percentage of nickel loading up to 20 or 23 wt.%. However, methane decomposition and CO disproportion are also responsible for coke formation over Ni/ZrO_2 catalysts with high nickel loadings and big particle size.

Basagiannis and Verykios [28-30] gave a great contribute about the catalyst influence on AASR reaction. Indeed, as a first approach, they performed AASR over Ni catalyst supported on La_2O_3-Al_2O_3 as well as over Al_2O_3 and La_2O_3 without Ni loading, following both transient and steady-state techniques. As a result, they demonstrated that acetic acid interacts strongly with Al_2O_3 and less with La_2O_3. Furthermore, they explained that Ni favours steam reforming reactions, retarding the carbon deposition rate onto the catalyst surface. They also analysed that carbon formation depends on the reaction temperature, steam/acetic acid feed ratio and catalyst composition. In detail, low T and low steam/acetic acid feed ratio makes possible carbon formation.

The same authors studied also the utilization of Ru (5wt%) dispersed on 15wt% MgO-Al_2O_3 during AASR reaction, demonstrating that it exhibits high activity and a good stability with time on stream. In particular with this kind of Ru-based catalyst, the reforming reactions and hydrogen production are favoured at relatively high T and low space velocities.

Other catalysts such as Pt, Pd, Rh, Ru and Ni were studied by Basagiannis and Verykios [28], analysing the influence of various supports among Al_2O_3, La_2O_3-Al_2O_3, MgO-Al_2O_3 and CeO_2-Al_2O_3. They highlighted that both Ni and Ru show high activity and selectivity toward hydrogen production and, in particular, Ru-based catalysts when supported onto La_2O_3-Al_2O_3 and MgO-Al_2O_3 present good long-term stability in dependence of time on stream. On the contrary, Ni-based catalysts did not show so high stability compared to Ru.

Takanabe et al. [23, 24] studied the effect of Pt/ZrO_2 catalyst used in AASR reaction, pointing out that the presence of ZrO_2 makes possible the production of acetone via aldol condensation (9), with consequent formation of coke deposits able to block the active sites of the catalyst. Nevertheless, Pt/ZrO_2 shows high activity toward AASR with the advantage of complete conversion of acetic acid and high hydrogen yield. Indeed, Pt plays an important role for AASR reaction, while ZrO_2 is necessary for steam activation. Unfortunately, these authors demonstrated that both conversion and hydrogen yield drop down rapidly in dependence of time on stream.

Also Iwasa et al. [16] studied the application of Pt-based catalysts to AASR reaction for hydrogen production, by analysing the activities of several metal catalysts among conventional metal oxides and transition metal-incorporated mesoporous smectite-like materials as supports. In particular, Iwasa and coworkers pointed out that Pt shows higher activity than Ni, Co and Fe among Al_2O_3 supported catalysts. Meanwhile, they stated that Al_2O_3 is more effective than ZrO_2 and SiO_2 as a support for Pt; Ni incorporated smectite (SM(Ni)) support is more effective than Fe and Co incorporated ones for Pt; SM(Ni) is also active if Pt is absent. Iwasa confirmed that the total activity order for converting acetic acid is Pt/Al_2O_3 > Pt/SM(Ni) > SM(Ni).

Another interesting study on AASR reaction was done by Wang et al. [27], who performed the reaction over a series of co-precipitated Co-Fe unsupported catalysts at relatively low T, highlighting that the higher the Co content the higher the catalytic activity. As a best result, Wang et al. found complete conversion and 96% of hydrogen yield using pure Co catalyst at T = 400 °C.

In conclusion of this short overview on the catalysts utilization in AASR reaction, we report the study of Bimbela et al. [9], who proposed the co-precipitated Ni–Al without specific support. In detail, three different Ni-catalysts were used with different metal loading percentage (23, 28 and 33%) and the best result was achieved with a Ni content of 28%, reaching a stable hydrogen yield at temperature not lower than 750 °C.

2. Membrane Reactor Technology

As reported in the previous paragraph, one of the main drawbacks of AASR reaction performed in FBRs is the carbon coke deposition, which is favoured at both low T and steam/acetic acid feed ratio [29, 32]. Nevertheless, Ni-based catalysts seem to be the most effective since they promote this reaction, depleting the rate of carbon deposition. Having in mind the production of high grade hydrogen for proton exchange membrane fuel cells (PEMFCs) supply, the FBRs utilization make possible the production of a reformed stream rich in hydrogen but still not useful for PEMFCs supply requirements due to the presence of other by-products such as CO_2 and CH_4 and, particularly, CO. Therefore, the hydrogen-rich stream coming out from a FBR needs further stages of separation/purification like water gas shift reactors (high and low temperature reactors), partial oxidation reactor and pressure swing adsorption step, all devoted to remove (or transform) the carbon monoxide, Figure 2a.

As an alternative to the conventional processes, dense Pd-based MRs make possible the combination of both reaction and hydrogen separation/purification in only one stage without needing further devices, Figure 2b.

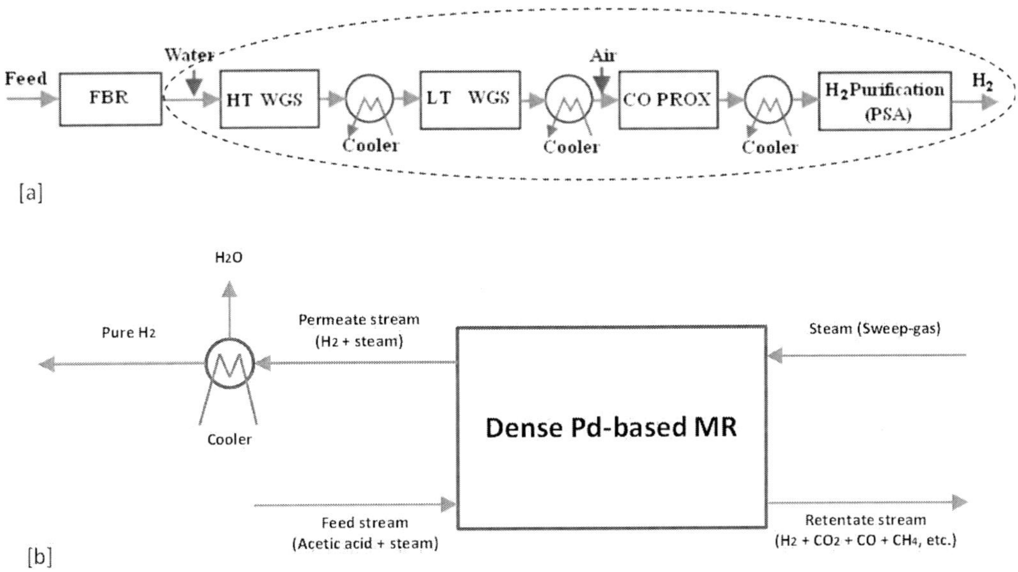

Figure 2. Conceptual scheme for high grade H_2 production from AASR reaction, [a] Conventional process; [b] Dense Pd-based MR (counter current flow configuration of sweep gas).

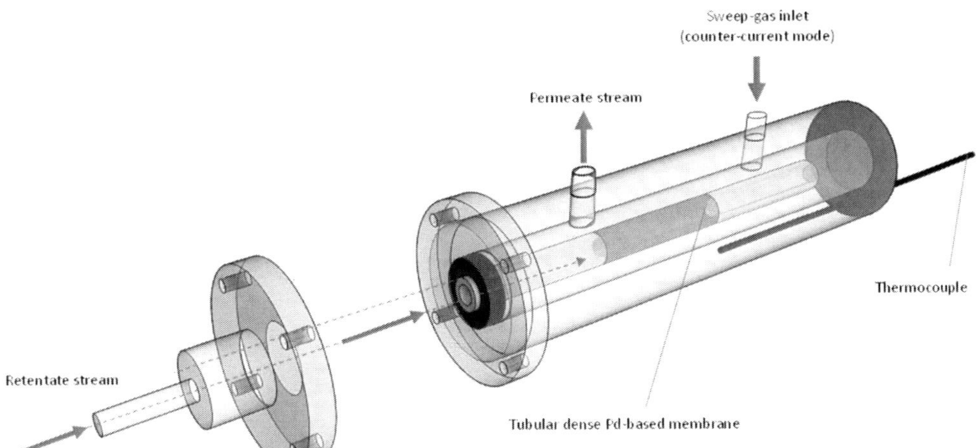

Figure 3. Scheme of a tubular shape MR housing a dense Pd-based membrane.

As a consequence, the conventional system (reformers + H_2-rich gas cleaning units) can be replaced by only one MR. Meanwhile, as shown in the conceptual scheme of Figure 2b, pure hydrogen can be collected in the permeate stream by using steam as a sweep gas (which can be easily separated via condensation), directly useful for PEMFC supply.

Otherwise, a vacuum pump could be also useful to strip the hydrogen removed from the reaction side for permeation through the Pd-based membrane.

Figure 3 and Figure 4 show two different MR modules in tubular shape. As illustrated in Figure 3, the tubular membrane is housed inside the module as in a tube in tube solution where the catalyst can be packed in the core of the membrane lumen or in the annulus of the shell.

Figure 4 shows another kind of tubular MR with an incorporated pre-heating zone.

However, the research on MR technology is a mature area and many researchers obtained relevant progress in the last years, particularly when it is devoted to hydrogen production from reforming reactions of both derived of fossil fuels and renewable sources [33-36].

2.1. Pd-Based MRs

In the last years, there has been a growing interest toward MRs, particularly when applied for hydrogen production processes owing to their continuous operation, modularity and reduced costs compared to the CRs. Special attention has been paid to the field of metal membranes such as Pd, Pd-alloy, Ti, Va, Mb, etc., particularly because of their hydrogen perm-selective characteristics even though among them, Pd and its alloys have been extensively studied.

As a general consideration, in a MR the effect that promotes the enhancement of the reaction conversion over the correspondent conventional FBR is the well-known "shift effect" [33]. As a special case, the utilization of dense and thin wall Pd-based membranes showing full hydrogen perm-selectivity with respect to all other gases allows the enhancement of the conversion due to the removal of hydrogen from the reaction side and, in the meanwhile, the recovery of high purity hydrogen. Therefore, a dense Pd-based MR allows higher conversions than a FBR operating at same experimental conditions, or the same conversion of a FBR, but operating at lower temperature with a consequent benefit due to the energy saving.

As stated above, dense Pd-based membranes have received a considerable attention owing to their full hydrogen perm-selectivity, even though in the temperature range of 0 – 700 °C other metals such as Nb, V and Ta show higher hydrogen permeability than Pd.

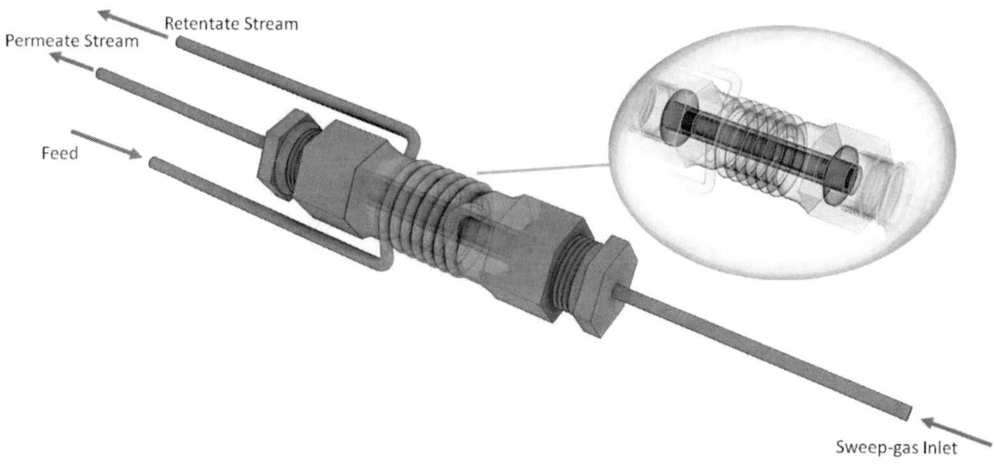

Figure 4. Scheme of a tubular MR with a preheating incorporated zone.

Nevertheless, these metals show greater surface resistance to hydrogen transport due to their easier oxidation.

The hydrogen molecular transport in Pd-based membranes follows the so called "solution/diffusion" mechanism, consisting of six different activated steps [37] as reported in Table 1.

Regarding the hydrogen permeation through dense Pd-based membranes, in the case of relatively low pressure, its rate-limiting step is assumed to be the diffusion [38]. Therefore, the hydrogen flux permeating through a Pd-based membrane can be expressed by the following equation:

$$J_{H_2} = Pe_{H_2} (p^n_{H_2,retentate} - p^n_{H_2,permeate})/\delta \qquad (12)$$

where J_{H2} is the hydrogen flux permeating through the membrane, Pe the hydrogen permeability, δ the membrane thickness, $p_{H2\text{-}retentate}$ and $p_{H2\text{-}permeate}$ the hydrogen partial pressures in the retentate and permeate zones, respectively, n (variable in the range 0.5 - 1) the dependence factor of the hydrogen flux on the hydrogen partial pressure. If the dense Pd-based membrane has a thickness higher than 5 μm, equation (12) becomes the Sieverts-Fick's law (13):

$$J_{H_2,Sieverts\text{-}Fick} = Pe_{H_2} \cdot (p^{0.5}_{H_2,retentate} - p^{0.5}_{H_2,permeate})/\delta \qquad (13)$$

However, the membrane thickness has a crucial role because as thin the membrane as high the hydrogen permeance. Nevertheless, thinner membranes can be affected by loss of mechanical resistance. Therefore, to propose dense Pd-based membranes in real applications, thicker membranes are necessary to guarantee adequate mechanical resistance and strength of the membrane, unfortunately increasing the cost due to higher amount of Pd.

2.2. Poisoning of Dense Pd-Based Membranes under MR Operation

Dense Pd-based membranes can show critical poisoning under operation. For example, pure Pd-membranes are affected by "embrittlement" phenomenon, which takes place below 300 °C and 2.0 MPa.

Table 1. Steps in solution/diffusion mechanism in Pd-based membranes

Hydrogen molecular transport in Pd-based membranes	
1st step	dissociation of molecular hydrogen at the gas/metal interface
2nd step	adsorption of the atomic hydrogen on membrane surface
3rd step	dissolution of atomic hydrogen into the palladium matrix
4th step	diffusion of atomic hydrogen through the membrane
5th step	re-combination of atomic hydrogen to form hydrogen molecules at the gas/metal interface
6th step	desorption of hydrogen molecules

To avoid this drawback, Pd is alloyed with other metals such as Ag, Cu, Y, etc. able to display its electron donating behaviour, allowing the competition for the electron holes filling within silver and hydrogen atoms [39].

Furthermore, palladium membrane surface suffers the contamination related to the presence of hydrogen sulphide, SO_2, arsenic, Hg vapour, thiophene, unsaturated hydrocarbons or chlorine carbon from organic materials, etc. In particular, the exposure to hydrogen sulphide can poison irreversibly Pd-coated membranes, which can be destroyed rapidly [40, 41].

As byproducts besides H_2, CO, CO_2 and CH_4 of various reforming reactions, some coke-precursors can be originated. Coke can cover the membrane surface, affecting negatively both the hydrogen permeation and the catalytic activity, reducing the whole reaction performance. The poisoning action is caused by the carbon atoms, which penetrate into the palladium lattice, making possible the membrane failure owing to the expansion of the palladium lattice [42].

Another contaminant for Pd-based membranes is constituted by CO, which is responsible for the decrease of the hydrogen permeation performances of the membrane, owing to the adsorbed CO that displaces the adsorbed hydrogen, blocking the hydrogen adsorption sites. This reduction is more evident at low temperature (lower than 150 °C) or at high CO feed concentration [43]. More than CO effect, the poisoning of water vapour on hydrogen permeating characteristic of Pd-based membranes is also effective owing to the water vapour dissociation/recombinative desorption, which contaminates the palladium surface with adsorbed oxygen [42].

3. Acetic Acid Steam Reforming in Both Conventional and Membrane Reactors

In this paragraph, AASR reaction is discussed paying particular attention to the most representative results achieved in FBRs in terms of conversion and hydrogen yield as well as the experimental conditions utilized such as temperature, space velocity, reaction pressure and feed molar ratio besides type of catalyst and reference, Table 2. Furthermore, in this table a few results are also shown regarding AASR in MRs. However, it is not possible to make a quantitative comparison among all the experimental data summarized in Table 2 owing to the different operating conditions adopted by each author.

Nevertheless, it is possible to observe from a qualitative point of view that most of the experimental results reported about FBRs are reached in the temperature range between 500 and 850 °C, with AcAc conversion ranging from 70 to 100%. Regarding to the application of AASR in MRs, to the best of our knowledge only two papers deal with MRs utilization [46, 47].

However, Table 2 depicts to the reader a scenario in which complete AcAc conversion is achievable from both FBRs and MRs even if the MRs operate at lower reaction temperature than FBRs with a consequent benefit in terms of energy saving.

Table 2 also shows that AASR has been studied using various catalysts, even if Ni-based are the most used, also considering that they are cheaper solution than noble metal-based catalysts such as Rh, Pt and Pd.

Table 2. AASR performance in terms of conversion and hydrogen yield at different experimental conditions for both CRs and MRs

T [°C]	Conversion [%]	$Yield_{H2}$ [%]	H_2O/Ac Ac [-]	p [bar]	Catalyst	GHSV [h^{-1}]	Kind of reactor	Reference
650	~ 90	0,118[b]	-	-	19%Ni/(Ni+Al)	28000	FBRs	Bimbela et al. [9]
500	~ 100	90	7,5/1	1	Co/Al_2O_3	8,3[c]	FBRs	Hu and Lu [10]
650	~ 100	-	2,4/1	1	Ni/Al_2O_3	10,2[c]	FBRs	Hu and Lu [11]
650	~ 100	92	3/1	1	Ni/CeO_2-ZrO_2	-	FBRs	Hu et al. [15]
700	~ 90	60	-	-	Pt/Al_2O_3	-	FBRs	Iwasa et al. [16]
400	~ 100	~ 90	7,5/1	1	16%Ni/ZrO_2	5,1[c]	FBRs	Li et al. [18]
650	~ 99	-	-	-	Ni/Al_2O_3	7000	FluBR	Medrano et al. [20]
800	~ 80	~ 80	9/1	1	Cu-Zu/Ca-Al	-	CR	Mohanty et al. [21]
700	~ 100	-	8,5/1	1	Ni-Co	-	CR	Pant et al. [22]
650	~ 80	62	1/1	-	$Ni/CeZrO_2$	-	CR	Thaicharoensutcharittha m et al. [25]
800	~ 80	-	-	-	0,5Rh/Al_2O_3	-	CR	Basagiannis and Verykios [28]
850	~ 70	-	-	-	1%Pd/Al_2O_3	-	CR	Basagiannis and Verykios [28]
730	92	-	3/1	1	$Ni/La_2O_3/Al_2O_3$	-	CR	Basagiannis and Verykios [29]
700	96	-	3/1	1	5%Ru/Al_2O_3	-	CR	Basagiannis and Verykios [44]
600	45-50	-	4/1	1	12wt% Ni/g-Al_2O_3	~ 5[a]	CR	An et al. [45]
400	100	-	10/1	1,5	Ru/Al_2O_3-Ni/Al_2O_3	-	dense Pd-Ag MR	Basile et al. [46]
400	-	56	10/2	1,5	Ni/Al_2O_3	-	dense Pd-Ag MR	Iulianelli et al. [47]

[a]WHSV.
[b]$Yield_{H2} = g_{H2}/g_{AcAc.}$
[c]LHSV.

As a general comment, the innovative use of dense Pd-based MRs allows various benefits such as to collect a highly concentrated hydrogen stream in the permeate side with a low CO content below 10 ppm. This ability makes also possible the combination of AASR in Pd-based MRs with PEMFCs.

Indeed, as well known, low-temperature (LT)-PEMFCs work at T < 100 °C (commonly ≈ 80 °C) and the CO tolerance for the supplied hydrogen stream is lower than 20 ppm. Therefore, the CO level of the reformed streams coming from FBRs is not adequate to directly supply LT-PEMFCs, needing further separation/purification of the hydrogen rich-streams.

On the contrary, as previously stated a dense Pd-Ag MR is able to supply a LT-PEMFC because of the full hydrogen perm-selectivity of the dense Pd-Ag membrane, which allows to collect hydrogen in the permeate stream with a CO content lower of the necessary requirements of the PEMFCs.

Conclusion and Future Trends

For future energy systems such as PEMFCs, hydrogen could play an important role as an energy carrier. Indeed, both industrial companies and academia are studying various ways to produce hydrogen in a more environmentally and economically attractive route. Steam reforming process and the exploitation of renewable sources are currently the most attractive way for producing hydrogen. An innovative approach discussed in this chapter has been the utilization of inorganic MRs as excellent candidates for the simultaneous production of hydrogen and its purification in a single unit. In particular, AASR process performed in both conventional and Pd-based MRs has been proposed, paying particular attention to the effect of dense Pd-based membranes on the reaction performances in terms of hydrogen yield and conversion, pointing out their special ability to collect highly concentrated hydrogen in the permeate side with a CO content lower than 10 ppm.

In summary, the future perspectives on performing AASR in inorganic MRs are: a) the development of low or not palladium-based membranes useful for AcAc reforming in MRs. This issue could represent a novel step for reducing the cost related to the palladium in the MRs; b) the scale-up of AASR in MRs could represent an important issue for larger application in industry; c) the development of low-cost Pd-based, resistant and defect-free membranes could represent a chance for realistic application of AASR reaction via MRs at industrial scale; d) great attention should be paid to evaluate the effective balance between benefits and drawbacks of applying MR technology to AASR process for producing hydrogen. In particular, further research should be devoted to analyse the increase of operating and capital costs caused by relatively high reaction pressure and temperature with the purpose of improving the hydrogen permeation driving force in hydrogen fully perm-selective MRs; e) more experimental studies on the combination and compatibility of new catalysts for AASR and MRs should be realized to better validate this process as a further way to produce hydrogen.

References

[1] S. Sorrell, J. Speirs, R. Bentley, A. Brandt, R. Miller, Global oil depletion: a review of the evidence, *En. Policy*, 38 (2010) 5290-5295.

[2] M. Hoogwijk, A. Faaij, R. van den Broek, G. Berndes, D. Gielen, W. Turkenburg, Exploration of the ranges of the global potential of biomass for energy, *Biomass Bioen.*, 25 (2003) 119-133.

[3] G. Fischer, L. Schrattenholzer, Global bioenergy potentials through 2050, *Biomass Bioen.*, 20 (2001) 151-159.

[4] A. Basile, A. Blasi, G. Fiorenza, A. Iulianelli, T. Longo, V. Calabrò, "Membrane and Membrane Reactor Technologies in the Treatment of Syngas Streams Produced from Gasification Processes", Ch. 4 pp. 139-174, In: *Gasification: Chemistry, Processes and Applications,* Michael D. Baker (Ed.), Nova Sci. Pub., 2011, ISBN: 978-1-61209-681-0.

[5] D. Wang, S. Czernik, D. Montane, M. Mann, E. Chornet, Biomass to hydrogen via fast pyrolysis and catalytic steam reforming of the pyrolysis oil or its fractions, *Ind. Eng. Chem. Res.*, 36 (1997) 1507-1518.

[6] G. W. Huber, S. Iborra, A. Corma, Synthesis of transportation fuels from biomass: chemistry, catalysts, and engineering, *Chem. Rev.*, 106 (2006) 4044-4098.

[7] Z. Qi, C. Jie, W. Tiejun, X. Ying, Review of biomass pyrolysis oil properties and upgrading research, *En. Conv. Manage.*, 48 (2007) 87-92.

[8] A. Oasmaa, D. Meier, Norms and standards for fast pyrolysis liquids 1. Round robin test, *J. Anal. Appl. Pyrolysis*, 73 (2005) 323-334.

[9] F. Bimbela, M. Oliva, J. Ruiz, L. García and J. Arauzo, Hydrogen production by catalytic steam reforming of acetic acid, a model compound of biomass pyrolysis liquids, *J. Analytical and Appl. Pyrolysis*, 79 (2007) 112-120.

[10] X. Hu, G. Lu, Comparative study of alumina-supported transition metal catalysts for hydrogen generation by steam reforming of acetic acid, *Appl. Catal. B: Environ.*, 99 (2010) 289-297.

[11] X. Hu, G. Lu, Bio-oil steam reforming, partial oxidation or oxidative steam reforming coupled with bio-oil dry reforming to eliminate CO_2 emission, *Int. J. Hydrogen En.*, 35 (2010) 7169-7176.

[12] X. Hu, G. Lu, Investigation of steam reforming of acetic acid to hydrogen over Ni–Co metal catalyst, *J. Molec. Catal. A: Chem.*, 261 (2007) 43-48.

[13] X. Hu, L. Zhang, G. Lu, Pruning of the surface species on Ni/Al_2O_3 catalyst to selective production of hydrogen via acetone and acetic acid steam reforming, *Appl. Catal. A: Gen.*, 427- 428 (2012) 49– 57.

[14] X. Hu, G. Lu, Acetic acid steam reforming to hydrogen over Co–Ce/Al_2O_3 and Co–La/ Al_2O_3 catalysts - The promotion effect of Ce and La addition, *Catal. Commun.*, 12 (2010) 50-53.

[15] R.-R. Hu, C.-F. Yan, X.-X. Zheng, H. Liu, Z.-Y. Zhou, Carbon deposition on Ni/ZrO_2-CeO_2 catalyst during steam reforming of acetic acid, *Int. J. Hydrogen En.*, DOI http://dx.doi.org/10.1016/j.ijhydene.2012.12.141.

[16] N. Iwasa, T. Yamane, M. Takei, J.-I. Ozaki, M. Arai, Hydrogen production by steam reforming of acetic acid: Comparison of conventional supported metal catalysts and metal-incorporated mesoporous smectite-like catalysts, *Int. J. Hydrogen En.*, 35 (2010) 110-117.

[17] N. Iwasa, T. Yamane, M. Arai, Influence of alkali metal modification and reaction conditions on the catalytic activity and stability of Ni containing smectite-type material for steam reforming of acetic acid, *Int. J. Hydrogen En.*, 36 (2011) 5904-5911.

[18] Z. Li, X. Hu, L. Zhang, S. Liu, G. Lu, Steam reforming of acetic acid over Ni/ZrO_2 catalysts: Effects of nickel loading and particle size on product distribution and coke formation, *Appl. Catal. A: Gen.*, 417-418 (2012) 281-289.

[19] Z. Li, X. Hu, L. Zhang, G. Lu, Renewable hydrogen production by a mild-temperature steam reforming of the model compound acetic acid derived from bio-oil, *J. Molec. Catal. A: Chem.*, 355 (2012) 123-133.

[20] J. A. Medrano, M. Oliva, J. Ruiz, L. Garcia, J. Arauzo, Catalytic steam reforming of acetic acid in a fluidized bed reactor with oxygen addition, *Int. J. Hydrogen En.*, 33 (2008) 4387-4396.

[21] P. Mohanty, M. Patel, K. K. Pant, Hydrogen production from steam reforming of acetic acid over Cu–Zn supported calcium aluminate, *Bioresou. Techn.*, 123 (2012) 558-565.

[22] K. K. Pant, P. Mohanty, S. Agarwala, A. K. Dalai, Steam reforming of acetic acid for hydrogen production over bifunctional Ni–Co catalysts, *Catal. Today*, DOI http://dx.doi.org/10.1016/j.cattod.2012.06.021.

[23] K. Takanabe, K.-I. Aika, K. Seshan, L. Lefferts, Catalyst deactivation during steam reforming of acetic acid over Pt/ZrO_2, *Chemical Eng. J.*, 120 (2006) 133-137.

[24] K. Takanabe, K.-I. Aika, K. Inazu, T. Baba, K. Seshan, L. Lefferts, Steam reforming of acetic acid as a biomass derived oxygenate: Bifunctional pathway for hydrogen formation over Pt/ZrO_2 catalysts, *J. Catal.*, 243 (2006) 263-269.

[25] S. Thaicharoensutcharittham, V. Meeyoo, B. Kitiyanan, P. Rangsunvigit, T. Rirksomboon, Hydrogen production by steam reforming of acetic acid over Ni-based catalysts, *Catal. Today*, 164 (2011) 257-261.

[26] Ekaterini Ch. Vagia, Angeliki A. Lemonidou, Investigations on the properties of ceria–zirconia-supported Ni and Rh catalysts and their performance in acetic acid steam reforming, *J. Catal.*, 269 (2010) 388-396.

[27] S. Wang, X. Li, L. Guo, Z. Luo, Experimental research on acetic acid steam reforming over Co-Fe catalysts and subsequent density functional theory studies, *Int. J. Hydrogen En.*, 37 (2012) 11122-11131.

[28] A. C. Basagiannis, X. E. Verykios, Catalytic steam reforming of acetic acid for hydrogen production, *Int. J. Hydrogen En.*, 32 (2007) 3343-3355.

[29] A. C. Basagiannis, X. E. Verykios, Reforming reactions of acetic acid on nickel catalysts over a wide temperature range, *Appl. Catal. A Gen*, 308 (2006) 182-193.

[30] A. C. Basagiannis, X. E. Verykios, Steam reforming of the aqueous fraction of bio-oil over structured $Ru/MgO/Al_2O_3$ catalysts, *Catal. Today*, 127 (2007) 256-264.

[31] D. Wang, S. Czernik, E. Chornet, Production of hydrogen from biomass by catalytic steam reforming of fast pyrolysis oils, *Energy Fuels*, 12 (1998) 19-24.

[32] S. Czernik, R. French, C. Feik, E. Chornet, Hydrogen by catalytic steam reforming of liquid by-products from biomass thermoconversion processes, *Ind. Eng. Chem. Res.*, 41 (2002) 4209-4215.

[33] A. Iulianelli, S. Liguori, T. Longo, A Basile, Inorganic membrane and membrane reactor technologies for hydrogen production, In: "Hydrogen Production: Prospects and Processes", Damon Robert Honery and Patrick Moriarty (Eds.) (Victoria, Australia), Nova Science Publishers, *Series: Energy Science, Engineering and Technology*, Ch. 12, pp. 377-398, ISBN: 978-1-62100-246-8 (2012).

[34] A. Iulianelli, S. Liguori, P. Pinacci, P. Morrone, A. Basile, Inorganic Membrane Reactor for Hydrogen Production through Bio-Ethanol Reforming Processes, In: *Ethanol: Production, Cellular Mechanisms and Health Impact*, R. I. Henry and B. P. Woods (Eds.); Nova Science Publishers, Ch. 3, ISBN: 978-1-62257-297-7 (2012).

[35] A. Iulianelli, S. Liguori, P. Morrone, A. Basile, Membrane and membrane reactor technologies for CO_x purification of gaseous streams, In: *Advance in Chemistry Research*, Nova Sci. Publ., Vol. 16, (2012).

[36] S. Smart, J. C. Diniz da Costa, S. Liu, J. M. Serra, A. Iulianelli, A. Basile, (2012). Porous ceramic membranes for membrane reactors. In: *Advance Membrane Reactors*, A. Basile, Editor. Chapter 8, (pp. 289-336) Woodhead Publishing Ltd: Cambridge, United Kingdom. DOI: 10.1533/9780857097330.2.298.

[37] W. J. Koros, G. K. Fleming, Membrane-based gas separation, *J. Membrane Sci.*, 83 (1993) 1-80.
[38] M. D. Dolan, N. C. Dave, A. Y. Ilyushechkin, L. D. Morpeth, K. G. McLennan, Composition and operation of hydrogen-selective amorphous alloy membranes, *J. Membrane Sci.*, 285 (2006) 30-55.
[39] J. Shu, B. P. A. Grandjean, A. Van Neste, S, Kaliaguine, Catalytic palladium-based membrane reactors: a review, *Can. J. Chem. Eng.*, 69 (1991) 1036-1060.
[40] D. J. Edlund, W. A. Pledger, Thermolysis of hydrogen sulfide in a metal-membrane reactor, *J. Membrane Sci.*, 77 (1993) 255-264.
[41] D. J. Edlund, W. A. Pledger, Catalytic platinum-based membrane reactor for removal of H_2S from natural gas streams, *J. Membrane Sci.*, 94 (1994) 111-119.
[42] B. A. McCool, Y. S. Lin, Nanostructured thin palladium-silver membranes: Effects of grain size on gas permeation properties, *J. Mater. Sci.*, 36 (2001) 3221-3227.
[43] A. Li, W. Liang, R. Hughes, The effect of carbon monoxide and steam on the hydrogen permeability of a Pd/stainless steel membrane, *J. Membrane Sci.*, 165 (2000) 135-141.
[44] A. C. Basagiannis, X. E. Verykios, Influence of the carrier on steam reforming of acetic acid over Ru-based catalysts, *Appl. Catal. B: Env.*, 82 (2008) 77-88.
[45] L. An, C. Dong, Y. Yang, J. Zhang, L. He, The influence of Ni loading on coke formation in steam reforming of acetic acid, *Renew. En.*, 36 (2011) 930-935.
[46] A. Basile, F. Gallucci, A. Iulianelli, F. Borgognoni, S. Tosti, Acetic acid steam reforming in a Pd-Ag membrane reactor: the effect of the catalytic bed pattern, *J. Membrane Sci.*, 311 (2008) 46-52.
[47] A. Iulianelli, T. Longo, A. Basile, CO-free hydrogen production by steam reforming of acetic acid carried out in a Pd-Ag membrane reactor: the effect of co-current and counter-current mode, *Int. J. Hydrogen En.*, 33 (2008) 4091-4096.

In: Acetic Acids
Editor: Angelo Basile

ISBN: 978-1-62948-217-0

Chapter 6

Acetic Acid in Micropaleontology: Its Efficiency in Disaggregating Carbonate Rocks and Recovering Calcareous and Siliceous Microfossils

Marlone H. Hünnig Bom*[*], *Karlos Guilherme Diemer Kochhann*, *Gislaine Bertoglio Rodrigues and Gerson Fauth
Instituto de Micropaleontologia (ITT FOSSIL), Universidade do Vale do Rio dos Sinos, São Leopoldo, Rio Grande do Sul, Brazil

Abstract

Micropaleontology provides essential information for dating sedimentary rocks and characterizing depositional environments, mainly based on calcareous (*e.g.*, ostracods, foraminifers) and siliceous (*e.g.*, radiolarians) microfossils. Several methodologies are used for breaking down rocks and recovering microfossils. The composition, degree of lithification, porosity and organic matter content are features controlling rock disaggregation and microfossils recovery. Therefore, pelitic and carbonate rocks react differently to the application of different methods. There are several methodologies for extracting microfossils from carbonate rocks, which take into account chemical and/or physical properties of the microfossils to be recovered and the rock. These disaggregation techniques are revised herein considering the immersion of carbonate rocks in acetic acid and ranging kinetic parameters. Their efficiency in recovering both calcareous and siliceous microfossils, from samples that otherwise would be considered barren, is demonstrated. Besides, acetic acid methods are simple and easy to work, also permitting reutilization of residues for further reactions by means of distillation.

[*] Corresponding author: *marloneb@unisinos.br*.

Introduction

Several techniques are used to disintegrate sedimentary rocks for micropaleontological purposes. Composition, degree of induration, porosity, and presence of organic matter are characteristics that play important roles in breaking down a rock sample and releasing its microfossiliferous content. Thus, siliciclastic (e.g., mudstones) and carbonate (e.g., limestones) rocks react distinctively when exposed to different chemical treatments.

Rock sampling (from outcrops or drilling holes) constitutes a crucial step of the analysis, having a significant influence on the results aimed. Samples of regular sizes are collected from different stratigraphic levels before being homogeneously fragmented (Figure 1). Size standardization of the particles is important to assure the efficiency of the method. Oxidation procedures using hydrogen peroxide (H_2O_2), usually applied to the disaggregation of rocks containing organic matter and/or low porosity/ permeability, are not efficient in disintegrating indurated carbonate rocks (characterized by their lower amount of organic matter).

Figure 1. Standard procedure applied to micropaleontological studies. Rock samples (B) are collected from outcrops (A) or drilling wells; they are then broken down in fragments of homogeneous sizes (C) before being disaggregated using physical/chemical treatments. The treated residue (D) is then analyzed for its microfossiliferous content (E). Ostracods, foraminifera and radiolarian are shown in E. The scale bars emphasize the different scales of observation that are taken into account for each step.

Problems also occur when target microfossils and the rock matrix have the same chemical composition, making necessary further precautions in order to recover well-preserved specimens (since the chemicals that attack the rock matrix are likely to damage the microfossils as well). Acetic acid (CH_3COOH) proved to be efficient in the recovery of microfossils, mainly calcareous forms, from carbonate rocks under a variable set of reaction conditions.

Bourdon (1956, 1962), followed by Lethiers and Crasquin-Soleau (1988), Tarsilli and Warne (1997), Crasquin-Soleau et al. (2005) and Rodrigues et al. (2012) successfully employed concentrated acetic acid (95% or more) for extracting calcareous microfossils from carbonate rocks. Some other reagents are also combined in the acetolisys: Thomas and Murney (1985) added a small amount of dehydrated copper sulphate on acetolisys; Moura et al. (1999) introduced diluted hydrochloric acid and hydrogen peroxide before acetic acid treatment. A weak solution of acetic acid dissolves the matrix of chalks and limestones, and is also efficient in extracting non-calcareous (especially phosphatized and silicified) fossils (Vodrázka, 2009).

In this chapter, a revision of the behavior of acetic acid treatments as applied to micropaleontology is presented, taking into account the reaction kinetics, the problem of corrosion and the reuse of the solvent.

The Acetic Acid Reaction

Several factors may influence the chemical process of disaggregating carbonate rocks, and consequently recovering calcareous microfossils. After choosing the most suitable reagent, the evaluation of kinetic parameters makes a crucial step in this procedure.

Among the several known methods, acetolysis may present the best results to disaggregate carbonate rocks. Its success may be related to the quantity of H^+ ions in the reaction. For instance, a strong acid as hydrochloric acid (HCl) is completely ionized in water; all HCl molecules ionize to form $H_3O^+{}_{(aq)}$ and $Cl^-{}_{(aq)}$ ions (Moore et al., 1998). Therefore, its reaction with carbonate rocks is extremely strong, damaging the microfossils (mainly the calcareous ones).

A weak acid, on the other hand, such as acetic acid, is only slightly ionized in water (less than 5% of the molecules are ionized) resulting in low H^+ concentration. Since the acid needs to penetrate into the rock pores, the reaction with calcium carbonate ($CaCO_3$) is slow and gradual, allowing more precise control of the reaction. Furthermore, the chemical reaction with calcium carbonate takes place between this salt and the H^+ ion, and not with the acid molecule itself (Krauskopf, 1989).

When the concentration of an acid varies, the ionization also presents some particularities: a strong acid, such as hydrochloric acid, dissolves calcium carbonate as presented in Equation 1:

$$CaCO_3 + 2H^+ \rightarrow Ca^{2+} + H_2O + CO_2 \qquad (1)$$

If this acid is in a diluted form (slightly ionized), the reaction (equation 2) would be somewhat different:

$$CaCO_3 + H^+ \rightarrow Ca^{2+} + HCO_3^- \quad (2)$$

The reaction speed is established by the concentration of the original ionic components. Since acetic acid produces low H^+, the reaction is slow ensuring good reaction control. Furthermore, when this acid is in its diluted form or when the sample contains moisture, the formation of calcium acetate crystals is favored besides the formation of hydronium ions (equation 3), which increase the dissolution of calcium carbonate.

$$CH_3COOH_{(aq)} + H_2O_{(l)} \rightarrow CH_3COO^- + H_3O^+_{(aq)} \quad (3)$$

Rodrigues et al. (2012) utilized 99.7% acetic acid, obtaining less dissociation of the rock sample than other works [Zolnaj (1979) used a concentration of 95% and Tarsilli and Warne (1997) applied a concentration of 96%]. Ideally, acetolysis should be applied in the purest form of the acid.

It is known that the most efficient way of speeding up a reaction is to increase its temperature. High temperatures, however, prevent accurate control of the reaction and may cause corrosion during acetolysis. Thus, mild temperatures and low or absent moisture during the reaction with acetic acid provide an effective procedure in extracting calcareous and siliceous microfossils from carbonate rocks.

In the micropaleontological research, to carry out an effective method, is as important as to assess how harmful this process might be to the environment and the damage it might cause in terms of waste generation. After breaking down carbonate rocks, acetic acid can be reused with the same efficiency (Crasquin-Soleau et al., 2005). Distillation can be used as a recycling procedure and, if possible, double-distillation. Before to start this process, it is important to filter the acid excess. Properties as the boiling point (118 °C) and pH (2.4 at 60.05 $g.L^{-1}$) (acetic acid safety data sheet, 1907/2006) should be considered during this process.

Acetolysis in Extracting Calcareous Microfossils

Foraminifera

Foraminifera are single-celled organisms (protists) that secrete organic (tectin) or mineral (mostly calcite, aragonite or silica) tests, or even tests composed of agglutinated material; these tests are normally built up in a series of chambers. Their stratigraphic record extends from the Cambrian (around 540 million years ago) to the Recent and they are planktonic and benthonic in habits, inhabiting from marine to fresh-water environments (Armstrong and Brasier, 2005).

Acetic acid has been used for recovering foraminifera from indurated limestones that would probably have to be studied by means of thin-sections of rocks otherwise (*e.g.*, Reolid and Herrero, 2004; Petrizzo et al., 2011; Coccioni et al., 2006), hampering accurate identifications of all faunal components (*e.g.*, Sliter, 1989). Most of the studies that have used acetic acid in recovering foraminifera applied the method of Lirer (2000), who adapted techniques previously adopted in the study of conodonts.

The technique consists in the digestion of calcilutites, marly calcilutites and calcarenites in cold, diluted (80%) acetic acid, for at least two hours (depending on the properties of the rocks, such as degree of induration and porosity), followed by a cleaning step in ultrasonic bath. That author considered that the reaction of acetic acid with carbonate rocks was slow but vigorous, and suggested that the acid was supposed to penetrate slowly within rock fragments destroying the links between the particles, despite the fact that the mechanism of reaction was unknown. Moura et al. (1999) also reported the use of diluted acetic acid (0.5 $mol.L^{-1}$) in recovering foraminiferal tests from carbonate rocks, although these authors did not evaluate corrosion effects on microfossils (just the number of specimens released per gram of treated rock). These authors suggested that the preferential reaction of the acid with the carbonate matrix is a consequence of its granular texture (increased contact surface) and contained impurities of the matrix. Reolid and Herrero (2004) adapted the method of Lirer (2004) adding Amine-O (a non-ionic industrial surfactant liquid) in an acetic acid solution and compared the degree of corrosion of this process with the acetolysis only. Thomas and Murney (1985) introduced in acetolysis a few grams of anhydrous copper sulphate, accelerating the acid attack of limestone and preventing dissolution effects on the calcareous calcareous microfossils. Kochhann (2012), however, used glacial acetic acid at room temperature, with indeterminate time of reaction [in accordance to experiment 12 of Rodrigues et al. (2012)], for recovering planktonic foraminifera from Albian indurated limestone samples that had been considered barren regarding foraminifera based on hydrogen peroxide methods. In Figure 2 it is presented foraminiferal specimens recovered with the use of glacial acetic acid under different temperatures and times of reaction. It is worth mentioning that the degree of corrosion seems to increase according to the reaction temperatures; while specimens recovered in room temperature show virtually no dissolution, specimens exposed during up to 48 h to acetic acid at 50 °C are characterized exclusively by casts (the foraminiferal test was already dissolved).

Ostracods

Ostracods are small bivalve crustaceans, mainly benthonic in habit, varying in size from 0.3 to 1.5 mm, and presenting a stratigraphic range that spans from the Ordovician (448 Ma) up to the present (Armstrong and Brasier, 2005). A quitinous-calcitic carapace secreted in equilibrium with the surrounding waters protects the soft body. They present a wide environmental distribution, inhabiting continental, transitional and marine aquatic environments.

Since ostracods present basically the same chemical composition as foraminifera (mostly calcium carbonate), the same methodologies applied in the extraction of the latter from carbonate rocks can be used with ostracods. Crasquin-Soleau et al. (2005) modified the acetolysis technique developed by Lethiers and Crasquin-Soleau (1988). The processing method consisted of heating a sample in acetic acid in a sand bath, from 12 h to three weeks (according to different microfacies) in high temperature.

They suggested that the washing step should be quick to control the acid attack and successive acetolysis should be done, if necessary. As the target microfossils have the same chemical composition of the rock matrix, corrosion can be a problem in this type of preparation.

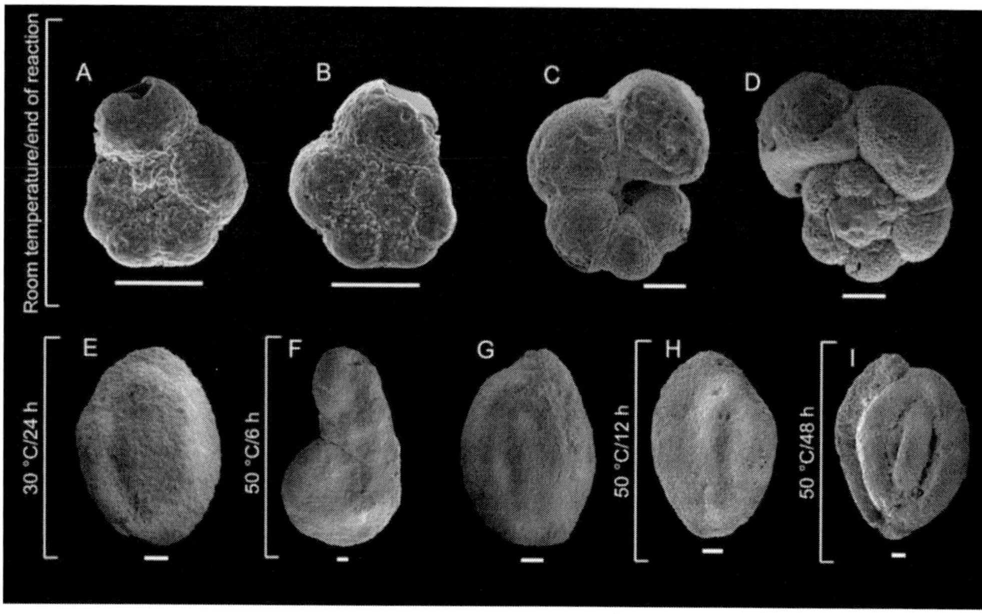

Figure 2. Foramifera recovered from carbonate rocks using acetic acid under different kinetic conditions. A-D: specimens recovered using acetic acid treatment at room temperature until a complete disaggregation of the rock was achieved, as suggested by Rodrigues et al. (2012) [provenance: Albian strata of DSDP (Deep Sea Drilling Program) Site 364]; E-I: specimens recovered using acetic acid heated at different temperatures with different periods of reaction (provenance: Albian strata from Potiguar Basin, Brazil); From the figure, it can be suggested that the longer the time of reaction in heated acetic acid the stronger the corrosion of foraminiferal tests (remaining only casts in Figure 2I). Scale bars: 50 μm.

A kinetic study was developed by Bom et al. (2011) aiming to assess the disaggregation of hard rock carbonatic samples for micropaleontological purposes. They suggested that ornamented carapaces are more resistant to dissolution than smooth ones (Figure 3). Andreau and Odin (2008), on the other hand, suggested that ostracod carapaces and valves recovered by means of acetolysis were weakened by corrosion and ornamentations were poorly preserved, although these authors did not explain the reaction conditions.

Rodrigues et al. (2012) tested different reagents in extracting ostracods from dolomitic carbonate rocks [$CaMg(CO_3)_2$], and the bests results were obtained with acetic acid. In the diluted form or when the sample contains moisture, disaggregation is faster and less efficient, forming calcium acetate (CH_3COOCa) and contributing to corrosion. Taking into account the studies discussed above, some details can play important roles in the application of acetolysis:

- Size standardization can contribute to better results. The size of rock fragments should be around 0.5 cm in diameter and then homogenized;
- Drying the sample in an oven prior to the chemical treatment prevents the formation of calcium acetate by reducing moisture;
- The temperature must be carefully controlled. A sand bath heats the sample in a more uniform way than a hot plate. The temperature should not exceed 100°C to avoid clay burning;
- The washing step should be quick to avoid corrosion due to further acid attack.

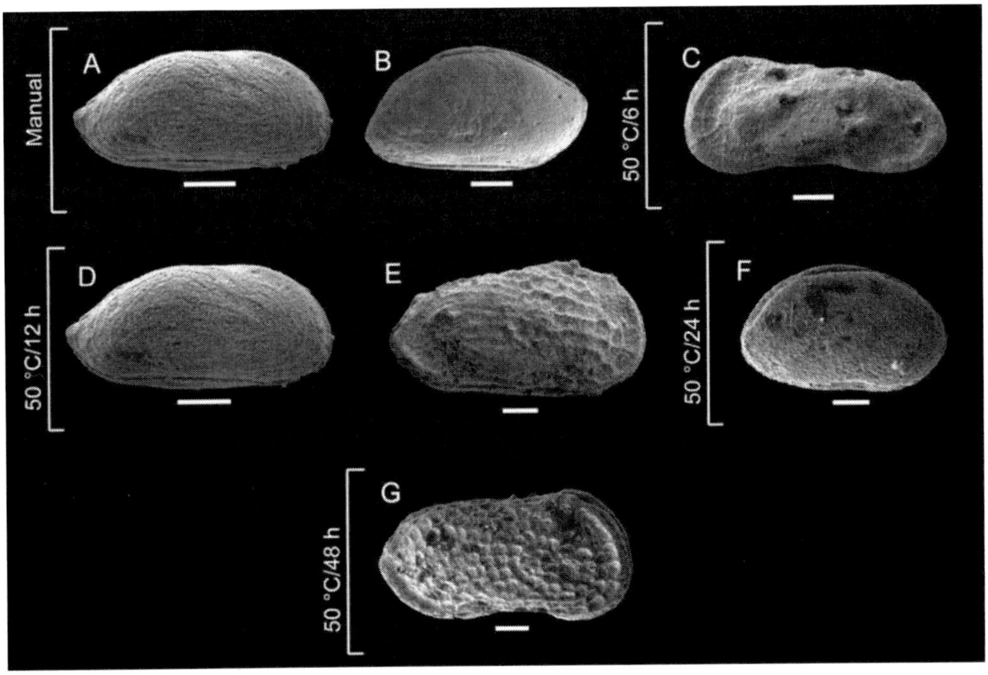

Figure 3. Ostracods recovered from carbonate rocks using acetic acid under different kinetic conditions. A-B: specimens recovered manually; C-G: specimens recovered using acetic acid heated at different temperatures with different periods of reaction (provenance: Albian strata from Potiguar Basin, Brazil). From the figure, it can be suggested that ostracods with conspicuous ornamentations are less prone to corrosion. Scale bars: 100 μm.

Acetolysis in Extracting Siliceous Microfossils

Radiolaria

Radiolarians are marine planktonic protists with a wide spacial and temporal distribution, inhabiting from low to high-latitude seas, from the Cambrian (around 540 Ma) to the present (*cf.*, De Wever et al., 2001; Kochhann, 2011). They secrete inner skeletons composed of amorphous opal (SiO_2) that can be preserved as fossils within marine sedimentary rocks. During diagenesis, the original amorphous opal of the skeleton can be turned into opal, opal-CT (cristobalite and/or tridymite) and finally into quartz, as the physical/chemical conditions change. Siliceous bioclasts (*e.g.*, radiolarian skeletons, diatom frustules, sponge spicules) can be easily recovered from limestones by means of hydrochloric acid preparation techniques (*cf.*, De Wever et al., 2001).

The hydronium concentration of HCl is quite high, characterizing a strong acid that reacts rapidly with calcium carbonate. Despite the efficiency of the hydrochloric acid in recovering siliceous microfossils from carbonate rocks, this is not the case as when epigeny (substitution of the bioclasts by other mineral phases) takes place. Once siliceous bioclasts are substituted by other mineral phases, mainly calcium carbonate, acetic acid treatments may provide the best results in isolating them from calcareous rocks.

Kariminia (2004), for instance, proposed a method for recovering radiolarian skeletons from limestones based on double immersion, by eight hours, of limestone fragments in acetic acid (CH_3COOH) at concentration of 75%. However, Kochhann et al. (2013) reported that Maastrichtian Radiolaria, probably replaced by calcium carbonate, recovered from DSDP Hole 999B using Kariminia's (2004) method showed worse preservation than specimens recovered from the same sample using the standard hydrogen peroxide (H_2O_2) method (taking into account samples in which the later method showed good results in the disaggregation of the rock). The reaction of the samples with H_2O_2, probably points out to the presence of organic matter in the clay fraction, since those rocks were described as clayey limestones (Sirgundsson et al., 1997). Thus, it seems reasonable that the application of glacial acetic acid in recovering siliceous bioclasts substituted by other mineral phases (mainly calcium carbonate), as previously suggested by Rodrigues et al. (2012) for calcareous microfossils, can provide better results than the immersion in the diluted form of the acid, avoiding corrosion due to a lower H^+ concentration, although this hypothesis still has to be tested. In Figure 4 radiolarian specimens replaced by calcium carbonate, recovered by means of the hydrogen peroxide and the diluted acetic acid methods, are compared.

It is also worth mentioning that techniques using hydrofluoric acid enable to recover siliceous radiolarian skeletons from siliceous sedimentary rocks such as cherts, in analogy to the way in which acetic acid treatments enable to recover radiolarian skeletons substituted by calcium carbonate from carbonate rocks (see discussion above). It occurs when the siliceous bioclasts are less soluble that the rock matrix, taking into account that the formers are compact microgranular structures that are likely to react as single crystals under controlled conditions (De Wever et al., 2001). Another reagent widely applied in the processes of extracting radiolarians from rocks is the deflocculating agent sodium hexametaphosphate [$(NaPO_3)_6$]. This substance is the active ingredient in the dispersant Calgon® (Hodgkinson, 1991).

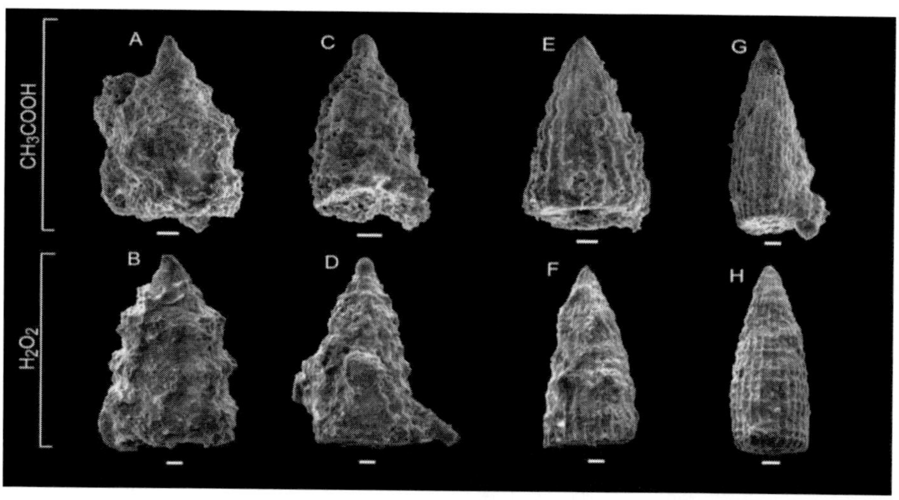

Figure 4. Radiolarians substituted by calcium carbonate recovered from Maastrichtian carbonate rocks of ODP (Ocean Drilling Program) Site 999. A-H: specimens recovered by means of the hydrogen peroxide and the diluted acetic acid (Kariminia, 2004) methods are compared. From the figure, it can be suggested that radiolarians recovered with acetic acid are more corroded, probably due to a higher H^+ concentration of the diluted form of the acid when it is compared to its pure form. Scale bars: 20 μm.

Depending on the chemical composition of the rock and the target radiolarian skeletons, one may apply the appropriate methodology to break the rock down, making compositions of some of the procedures exposed above. In some cases, processes with different stages of acidification can be applied (e.g., Hollis and Hanson, 1991).

Conclusion

The usefulness of acetic acid in recovering calcareous and siliceous microfossils from carbonate rocks is highlighted herein. Since it is a weak acid and only slightly ionized in water (less than 5% of the molecules are ionized), resulting in low H^+ concentration, it has to penetrate within the rock pores. In this way, the reaction with calcium carbonate is slow and gradual, allowing a precise control of the reaction. The methods revised herein are efficient in recovering calcareous microfossils (e.g., ostracods, foraminifers), from samples that otherwise would be considered barren, and siliceous microfossils (e.g., radiolarians), mainly when the original biogenic silica is substituted by calcium carbonate. Once one takes the precautions detailed in the present review, acetic acid methods are easy to work, also permitting reutilization by distillation of residues for further reactions.

References

Acetic Acid; MSDS No. 1907/2006; Sigma-Aldrich Brasil Ltda: São Paulo, Apr. 21, 2012.

Andreu, B., Odin, G. S. New micorpaleontological studies on the type section of the Campanian-Maastrichtian at Tercis (SW France): new ostracodes obtaining using acetolysis. *Notebooks on Geology* 2008, CG2008_A02, 1-14.

Armstrong, H. A., Brasier, M. D. *Microfossils*, 2nd edition; Blackwell Publishing: Oxford, 2005.

Bom, M. H. H., Bergue, C. T., Rodrigues, G. B., Fauth, G. Aplicação da acetólise na recuperação de microfósseis calcários em rochas carbonáticas do Lajedo Soledade (Formação Jandaíra, Bacia Potiguar). In: *Paleontologia: Cenários da Vida*; Carvalho, I. S., Srivastava, N. K., Strohschoen Jr., O., Lana, C. C., Eds.; Editora Interciência: Rio de Janeiro, 2011; Vol. 3, pp. 117-122.

Bourdon, M. Utilisation de l'acide acétique dans desegregation des roches dures. *Revue de l'Instituit Français du Pétrole* 1957, *12*, 14-15.

Bourdon, M. Methode de degagement des microfossiles par acetolyse a chaud. *Compte Rendus Sommelier de Societe Geologique de France* 1962, 267-268.

Coccioni, R., Luciani, V., Marsili, A. Cretaceous oceanic anoxic event and radially elongated chambered planktonic foraminifera: Paleoecological and paleoceanographic implications. *Palaeogeography, Palaeoclimatology, Palaeoecology* 2006, 235, 66-92.

Crasquin-Soleau, S., Vaslet, D., Nindre, Y. M. Ostracods as markers of the Permian/Triassic boundary in the Khuff Formation of Saudi Arabia. *Paleontology* 2005, 48, 853-868.

De Wever, P., Dumitrica, P., Caulet, J. P., Nigrini, C., Caridroit, M. *Radiolarians in the sedimentary record*; Gordon and Breach Science Publisher: New York, 2001.

Hollis, C. J., Hanson, J. A. Well-preserved late Paleocene Radiolaria from the Tangihua Complex, Camp Bay, eastern Northland. *Tane* 1991, 33, 65-76.

Hodgkinson, R. L. Microfossil processing: a damage report. *Micropaleontology* 1991, 37, 320-326.

Kariminia, S. M. Extraction of calcified Radiolaria and other calcified microfossils from micritic limestones utilizing acetic acid. *Micropaleontology* 2004, 50, 301-306.

Kochhann, K. G. D. Radiolários: estado do conhecimento e aplicações às geociências. *Terrae Didatica* 2011, 7, 18-28.

Kochhann, K. G. D. Foraminíferos planctônicos e bentônicos do intervalo Aptiano-Albiano do DSDP Site 364 (Bacia De Kwanza): taxonomia, bioestratigrafia, paleoecologia e implicações paleoceanográficas. M.Sc. Dissertation, Universidade do Vale do Rio dos Sinos, November 2012.

Kochhann, K. G. D., Baeker-Fauth, S., Fauth, G. Late Maastrichtian Radiolaria from ODP Leg 165 – Site 999B, Colombian Basin. *Geodiversitas* 2013, 35, 5-20.

Krauskopf, K. B. *Introduction to Geochemistry*, 2nd edition; McGraw-Hill Book Co.: New York, 1989.

Lethiers, F., Crasquin-Soleau, S. Comment extraire les microfossiles à tests calcitiques des roches calcaires dures. *Revue de Micropaléontologie* 1988, 31, 56-61.

Lirer, F. A new technique for retrieving calcareous microfossils from lithified lime deposits. *Micropaleontology* 2000, 46, 365-369.

Moore, J. W., Stanitski, C. L., Wood, J. L., Kotz, J. C., Joestern, M. D. *The Chemical World: Concepts and Applications*, 2nd edition; Saunders College Publisher: Orlando, 1998.

Moura, J. C., Rios-Netto, A. M., Wanderley, M. D., Souza, F. P. Using acids to extract calcareous microfossils from carbonate rocks. *Micropaleontology* 1999, 45, 429-436.

Petrizzo, M. R., Falzoni, F., Premoli Silva, I. Identification of the base of the lower-to-middle Campanian Globotruncana ventricosa Zone: Comments on reliability and global correlations. *Cretaceous Research* 2011, 32, 387-405.

Reolid, M., Herrero, C. Evaluation of methods for retrieving foraminifera from indurated carbonates: application to the Jurassic spongiolithic limestone lithofacies of the Prebetic Zone (South Spain). *Micropaleontology* 2004, 50, 307-312.

Rodrigues, G. B., Bom, M. H., Fauth, G. Recovery of ostracods in Cretaceous dolomitic carbonate: The efficiency of acetolysis. *Marine Micropaleontology* 2012, 92–93, 81–86.

Sliter, W. V. Biostratigraphic zonation for Cretaceous planktonic foraminifers examined in thin section. *Journal of Foraminiferal Research* 1989, 19, 1-19.

Sigurdsson, H., Leckie, R. M., Acton, G. D., Abrams, L. J., Bralower, T. J., Carey, S. N., Chaisson, W. P., Cotillon, P., Cunningham, A. D., D'hondt, S. L., Droxler, A. W., Galbrun, B., Gonzalez, J., Haug, G., Kameo, K., King, J., Lind, I. L., Louvel, V., Lyons, T. W., Murray, R. W., Mutti, M., Myers, G., Pearce, R. B., Pearson, D. G., Peterson, L. C., Röhl, U. *Proceedings of the Ocean Drilling Program.*; Texas A and M University: College Station, 1997; Vol. 165.

Tarsilli, A., Warne, M. T. An improved technique for extracting calcareous microfossils from Paleozoic limestones. *Alcheringa* 1997, 21, 57-64.

Thomas, F. C., Murney, M. G. Techniques for extraction of foraminifers and ostracods from sediment samples. *Canadian Technical Report of Hydrology and Ocean Sciences* 1985, 54, 1-24.

Vodrázka, R. A new method for the extraction of macrofossils from calcareous rocks using sulphuric acid. *Palaeontology* 2009, 52, 187-192.

Zolnaj, S. Carbonate microfossil separation from limestone. In: *Taxonomy, biostratigraphy and distribution of ostracodes*; In: Krstic, N., Ed.; Serbian Geological Society: Belgrade, 1979; pp. 265-269.

In: Acetic Acids
Editor: Angelo Basile

ISBN: 978-1-62948-217-0

Chapter 7

Acetic Acid in the Food Industry

***Đurđica Ačkar**, *Drago Šubarić*,**
Antun Jozinović and Jurislav Babić
Faculty of Food Technology Osijek, Osijek, Croatia

Abstract

Acetic acid has been known to humanity for many centuries. It was first discovered in the form of vinegar, when wine got spoiled by the action of bacteria. Ancient Romans cooked soured wine to produce highly sweet syrup.

Since then, the importance of acetic acid in food preparation/production steadily grew. Acetic acid used in food preparation/production can only be produced by natural process – aerobic fermentation, through action of acetic bacteria. In USA it is GRAS additive and in EU it is designated as E 260. It has no adverse effects and no dietary restrictions.

Today, acetic acid is not used only as vinegar in salad preparation. It is used in the preservation of food as a natural preservative and antimicrobial agent not only in vegetable pickling, but as an active component of edible films and in hurdle technology of preservation. It is added to mayonnaise to deactivate *Salmonella.* In addition, it is acidulant, gives a characteristic flavour profile to food, can be used for microbial decontamination of meat and as mild descaling agent in food industry. Research on acetic acid applications in the accelerated ripening of dates is also conducted.

Indole-3-acetic acid plays a role in banana ripening (phytohormone) and its degradation to 2-aminoacetophenon during wine production can give untypical aging off-flavour in wine. It can be used in grey mould control in harvested apple fruit.

Acetic acid and its anhydride are widely used in the production of modified starches, so called starch acetates as well as dual-modified starches.

This chapter will summarise current knowledge on acetic acid and its derivatives application in the food industry.

Keywords: Acetic acid, production, application in food

* Corresponding author: Đurđica Ačkar, dackar@ptfos.hr.

Introduction

Acetic acid has been known to humanity for many centuries. It was first discovered in the form of vinegar, when wine got spoiled by action of bacteria. Although ancient civilizations didn't know that "soured wine or beer" was result of bacterial fermentation or that acetic acid was formed, they soon discovered that these "spoiled" liquids could be used in food preservation.

In the East, production of vinegar from palm wine was known in 5000 BC. In Ancient Rome vinegar was used in food preservation, often combined with salt, wine or honey (Luck & Jager, 1997). Since no sugar was available, Ancient Romans prepared "sugar of lead" (lead acetate) from wine vinegar which contributed to lead poisoning (Retief & Cilliers, 2006).

The first industrial production of vinegar was noted in France in 14th century (Luck & Jager, 1997). This production process was patented in 16th century as Orleans process, named after town in France and it implied wine addition to "mother vinegar" and maturation in wooden barrels.

In 1864 Pasteur discovered that *Mycoderma aceti* caused acidification of wine to vinegar and improved original Orleans process (Tesfaye et al., 2002).

Nowadays, vinegars are not produced solely from wine. According to FAO/WHO Food Standards programme proposition (2000), vinegar is „a liquid fit for human consumption, produced exclusively from suitable products containing starch and/or sugars, by the process of double fermentation, first alcoholic and then acetous, and contains a specified amount of acetic acid".

Vinegar Types

Although acetic acid can be produced by destructive distillation of wood and catalytic oxidation of ethanol or acetaldehyde, for human consumption acetic acid can only be produced by biocatalytic bacterial oxidation (Blech, 2009).

Acetic acid bacteria are rod shaped, aerobic, Gram negative, catalase positive microorganisms, classified in 12 genera and 59 species in *Acetobacteraceae* family. Genera *Acetobacter* and *Gluconacetobacter* oxidise ethanol to acetic acid by action of alcohol dehydrogenase and aldehyde dehydrogenase and these genera are involved in vinegar production (Andres-Barrao et al., 2012).

According to Blech (2009), three major types of vinegar are produced: wine vinegar, cider vinegar and white vinegar. These types of vinegar differ in taste due to nonacetic acid portion of the product. According to Croatian legislative, wine and cider vinegars contain minimally 60 g acetic acid/L and max 1.0% vol of alcohol. FAO/WHO Proposed Regional Standard for vinegar (2000) requires not less than 60 g/L acetic acid for wine vinegar and not less than 50 g/L for other vinegar types, and not more than the amount detainable through biological fermentation. According to this standard, wine vinegar can contain up to 0.5% (v/v) residual alcohol, while other types can contain up to 1.0% (v/v) residual alcohol.

Wine Vinegar

Wine vinegar can be produced by two methods: traditional (slow) process and submerged (quick) process. Traditional process is surface culture fermentation. Bacteria are placed on the surface of the liquid and are in direct contact with air. This is static method which requires a long period of time, but results in high-quality product and is employed in production of traditional and selected vinegars (Tesfaye et al., 2002).

Submerged fermentation was first described by Hromatka & Ebner in 1949. Here, bacteria are submerged in the liquid and oxygen is constantly supplied in very fine bubbles generated by special turbines to prevent excessive evaporation of acetic acid and ethanol (Schleputz, Gerhards & Buchs, 2013). The main advantage of this method is short production time: 24 – 48 h (Sengun & Karabiyikli, 2011). The process can be batch, semi-continuous or continuous and most often used columns are Bubble column fermenter and Frings acetator (Tesfaye et al., 2002). Process is most efficient at 30 °C and temperature increase by 2 – 3 °C results in significant yield reduction. In addition, to increase yield, evaporation of ethanol, acetaldehyde and acetic acid has to be reduced to minimum, contact of substrate with air prolonged and oxygen availability assured (Tesfaye et al., 2002).

Jerez or sherry wine vinegar is traditional Spain product, produced exclusively from sherry wine made from Palomino, Pedro Ximenez and Muscat grapes grown in vineyards registered in "Jerez-Xeres-Sherry" Denomination. It can be produced either by Orleans or by submerged method and is aged in so-called "solera" system, series of casks arranged in three to eight steps. Substrate is introduced on the top of the system and final product (one-third of total volume) is withdrawn at the bottom, vinegar from stage two is filled in stage 1 barrels and vinegar from stage 3 to stage 2 barrels to homogenise product through the whole system. Sherry vinegar ages minimum 6 months. Vinegar aged for 2 years is Reserva Sherry vinegar and Gran Reserva vinegar is aged for more than 2 years (Tesfaye et al., 2002).

Unlike other wine vinegars, Jerez vinegars contain significant amounts of amino acids. Their content increases by double from the beginning of production to solera stage. During aging, polyphenol content increases, with predominant gallic acid (Solieri & Giudici, 2009).

Balsamic vinegar (Aceto Balsamico di Modena) is Italian traditional product, produced from concentrated grape must. In traditional process, must is first cooked at boiling temperature, skimmed to remove dispersed solids and coagulated proteins, and then temperature is maintained at 80 – 90 °C for several hours. Next step is two-stage fermentation. In the first stage, yeasts convert sugar to alcohol which is further converted to acetic acid by *Gluconobacter*. Traditional method requires storage in wood barrels for minimum 12 years and up to 25 years. Hence, longer the maturation, darker brown colour is formed (Solieri & Giudici, 2009).

In industrial conditions, concentrated must is mixed with acetic acid and matured in wooden barrels. Wood shavings can be added to speed up the process (Tesfaye et al., 2002) and grape juice and addition of commercial cultures reduce production costs (Blech, 2009).

Traditionally produced balsamic vinegar, in addition to acetic acid, contains significant amounts of tartaric, gluconic, malic and succinic acid. This is why balsamic vinegar is less pungent than other vinegar types and has characteristic aroma. In addition, balsamic vinegar contains antioxidants: polyphenols, mainly phenolic acids, catechines, polymeric procyanidines and flavonols, and melanoidines (Solieri & Giudici, 2009).

Cider Vinegar (Solieri & Giudici, 2009)

Cider or apple vinegar is produced from apple juice by double fermentation (alcoholic and acetic). Traditionally, alcoholic and acid fermentation are carried out in the same barrel, covered with cloth and placed in a warm, damp place. The process takes 5 – 6 month and has low yields. Therefore, nowadays, traditional method is largely replaced with submerged.

Apples are crushed and juice is extracted by pressing grated apples. Alcoholic fermentation of the juice is carried out by *Saccharomyces cerevisiae*. At the end of the process yeast cells and fruit pulp settle to the bottom and fermented liquid is generally separated by centrifugation. Before acetic acid bacteria inoculation, alcohol content is set to 6 – 7%. After acetification, cider vinegar is aged in wooden or stainless steel tanks for 1 to 2 months. During maturation, harsh flavour is lost and fruity aroma is formed, with colour intensity improvement.

Aged vinegar is clarified by sedimentation or ultrafiltration and pasteurised to avoid spoilage. Characteristic of cider vinegars is the presence of significant sorbitol content and phloridzin.

During storage, post-bottling haze and darkening can occur due to iron, copper, oxygen, procyanidin and oxidised ascorbic acid presence. To prevent this, SO_2, ascorbic acid, pectin, citric acid or potassium ferrocyanide can be added.

Another cider vinegar defect is growth of “mother of vinegar”, which occurs when *Gluconacetobacter xylinus* forms cellulosic film. During production and storage of vinegar, at the surface of the liquid nematodes *Turbatrix aceti* may grow. They can be eliminated by pasteurisation, filtration or inhibited by addition of SO_2.

Malt and White (Spirit) Vinegar

Several types of malt vinegar are available: distilled, light and dark malt vinegar.

Malt vinegar is pale straw-coloured liquid with strong acetous flavour derived from malted barley by double fermentation without intermediate distillation. Dark malt vinegar is produced by addition of barley extract or caramel to the malt vinegar (Solieri & Giudici, 2009).

First stages of malt vinegar production are similar to beer production. Acetification of alcoholic liquid is performed by acetifiers or acetators. Rough stock vinegar is aged for minimum 3 months (Solieri & Giudici, 2009). Aged vinegar is clarified and total acidity and colour are corrected according to consumers’ preference.

Generally, white vinegar is produced from ethanol derived through fermentation of carbohydrates or from petrochemical sources. Traditionally, this vinegar type is produced from malted barley (Blech, 2009). Distilled malt vinegar is colourless, mildly aromatic and sharper in aroma than malt vinegar. It is produced by distillation of malt vinegar before maturation and therefore contains only volatile constituents.

Other Vinegar Types

Rice vinegars are produced from rice wine (sake) and traditionally consumed in China, Japan and Korea. There are three types of rice vinegar. Amber vinegar (komesu) is used in sweet-and-sour dishes, red is used in steamed and boiled crab, and black (kurosu) is consumed as table condiment or healthy drink (Solieri & Giudici, 2009).

Komesu is colourless Japanese vinegar produced from polished rice. Saccharification of rice is conducted by *Aspergillus oryzae, Rhizopus, Mucor* and *Monascus*. Alcohol fermentation is performed by sake yeast *Saccharomyces cerevisiae*. Traditionally, moromi (liquid containing alcohol mixed with vinegar wash) is fermented by acetic bacteria in tightly covered containers for one month (Solieri & Giudici, 2009).

Vinegar produced from unpolished rice is kurosu. It is famous for its health benefiting effects (aid for hypertension; cancer prevention) due to relatively high content of amino acids, vitamins and soluble metal ions. Production process is very similar to komesu production.

Sake-lees vinegar (kasuzu) is fermented from sake production waste material, so called sake lees extracts. Distinguishing characteristic of this vinegar type is that alcohol fermentation is not performed. Sake lees are aged for 2 years, mixed with alcohol and acidified with vinegar to make moromi, which is fermented by traditional static method (Solieri & Giudici, 2009).

Red rice vinegar has characteristic red colour resulting from the red mould. Rice is steamed, saccharified and fermented with red koji. Obtained moromi is mixed with one year vinegar batch and aged for 3 years (Solieri & Giudici, 2009).

Zhenjiang aromatic vinegar is traditional Chinese vinegar from glutinous rice. After saccharification and alcoholic fermentation in semi-solid state, acetic fermentation is conducted in solid state by mixing alcohol mash with wheat bran and rice hulls. This stage lasts for app. 20 days and temperature is maintained in 40 – 46 °C interval, with humidity maintained at 60 – 70%. Fermentation culture of 7th day is used as starter for the next batch. Aging of the fermented liquid lasts for over three month (Xu et al., 2011). It contains ligustrazin, which has function in lowering blood pressure (Chen & Chen, 2009).

Shanxi vinegar is traditional Chinese product produced by spontaneous solid-state fermentation of several cereal types including sorghum and starter Daqu (60% raw material made of barley and pea) with addition of wheat bran and millet chaff as filling and supplementary material. In addition to raw material, characteristic of this vinegar is combined action of moulds, yeast, lactic and acetic bacteria which results in unique flavour and aroma (Wu et al., 2012).

Whey vinegar is produced from whey by dual fermentation process. Alcoholic fermentation is performed in microaerophilic conditions by *Kluyveromyces maxianus*, and *Gluconacetobacter liquefaciens* or *Acetobacter pasteurianus* convert ethanol to acetic acid.

Coconut vinegar is cloudy white liquid with sharp acidic taste and yeasty flavour made from sap of coconut palm (Solieri & Giudici, 2009).

Cane vinegar is dark yellow to golden brow vinegar with mellow flavour made from sugar cane.

In Africa, sorghum, mango, palm, banana, date, bamboo and honey vinegar are produced. In Taiwan pineapple vinegar is produced by mixing pineapple juice with rice vinegar (Solieri & Giudici, 2009).

Antimicrobial Activity of Acetic Acid

Weak organic acids are widely used to prevent microbial spoilage of foods due to extension of lag phases, which is their main inhibitory effect (Fernandez, 2005). According to literature data, inhibitory effects are result of action on nutrient uptake and energy metabolism, action as uncouplers and accumulation effect (Theron & Lues, 2011).

Namely, undissociated organic acids are lipophilic and under acidic conditions freely cross the microbial membrane to the cytoplasm, where they cause acidification and dissipation of the membrane pH gradient. In addition, the acids can accumulate in the cell and destabilise cell proteins, inhibit metabolic actions and cause accumulation of toxic anions (Stratford et al., 2009; Theron & Lues, 2011). However, this classical "weak-acid theory" does not fully describe antimicrobial action of organic acids and additional research is needed to elucidate why acids with equal pK_a value have different toxicity to microorganisms (Stratford et al., 2009).

It has been established that acetic acid has inhibitory effect on *Aspegillus niger* and *A. flavus* due to rapid falls of pH_i (Stratford et al., 2009; Leon Pelaez et al., 2012) and slows down growth of *Lysteria monocytogenes* due to decrease of pH (George et al., 1996) even in small concentration of 0.2% (Vermeulen et al., 2007).

Greenacre & Brocklehurst (2006) investigated acid tolerance response of *Salmonella typhimurium* to acetic acid. They found that acetic acid-tolerance is developed by bacteria in mildly acidic conditions and that these bacteria resist NaCl osmotic stress too.

Lee & Kang (2009) examined combined effect of heat, acetic acid and NaCl treatment on survival of *Escherichia coli* O157:H7. They found that heat synergistically acted with acetic acid, while NaCl had antagonistic effect. Namely, salt addition increased pH value and reduced effectiveness of acid.

Salmonella susceptibility to acetic acid is affected by the physiological state of bacteria with highest sensitivity of *S. bareilly* and the lowest of *S. stanley*. Synergistic effect of acetic acid with thymol and carvacrol on *Salmonella typhimurium* was also established (Mani Lopez et al., 2012).

Acetic Acid Application in Food Industry

Vegetable Processing

Vinegar has extended application as condiment in salads and dressings due to unique flavour. However, more important acetic acid application in food is based on its antimicrobial activity. It is used as preservative against fungi and bacteria and in European laws it is marked as E 260 in category of "other additives" since it can be used as acidity regulator as well (EUFIC). It has no side effects and in the USA it is recognised as GRAS (Generally Recognised as Safe) additive with unlimited acceptable daily intake (Mani Lopez et al., 2012).

Most often acetic acid is applied in marinated vegetable production. Preservation of vegetables in brine of vinegar and salt is based on anabiosis principle: microbial activity is limited or suppressed due to combined action of acidification and osmoanabiotic action of salt. To improve preservation effect and/or taste, sugar and/or oil can be added to the brine solution. In order to determine if acetic acid is successfully applied to preserve the product without heat or other preservative application, Preservation Index (PI) is used:

$$PI = (\text{total acetic acidity} \times 100)/(100 - \text{total solids})\ [\%].$$

If the preservation index (PI) is equal or higher than 3.6%, product is reasonably safe from spoilage (Ranken, Baker & Kill, 1997).

Vegetable intended for marinating has to be healthy, undamaged and properly prepared.

Pickled cucumbers are calibrated according to size, washed, dosed into glass jars and brine solution is added. Different condiments (e.g. dill) can be added before filling jars with the solution. Jars are sealed and pasteurised (Lovrić & Piližota, 1994).

Peppers can also be pickled in brine. Different sorts of peppers can be pickled both whole and shredded. Before pickling, peppers are blanched at 80 °C for 2 – 5 minutes, cooled and filled into jars. Brine usually contains up to 3% salt and 3% acetic acid (Lovrić & Piližota, 1994).

Red beet is washed, cooked (60 to 90 min at 90 – 95 °C), peeled and sliced before pickling. After brine addition, product has to be pasteurised (Lovrić & Piližota, 1994).

Vinegar is common ingredient of *ketchup,* heterogeneous spiced product produced from tomato concentrate, puree or paste (Sahin & Ozdemir, 2004; Mert, 2012). Vinegar concentration in this product can be as low as 0.1 – 0.2% and to prevent spoilage by lactobacilli, ketchup has to be pasteurised.

Combination of different approaches to food preservation is commonly known as hurdle technology. By intelligent combination of "hurdles" microbial safety of food is preserved with minimum intervention in organoleptic and nutritional quality, and economic viability (Leistner, 2000). Most important "hurdles" in food technology are temperature (heating or cooling), water activity, acidity, redox potential, preservatives and competitive microorganisms. Same hurdle can have multiple effects, for instance, acetic acid is preservative which changes pH of the food and influences organoleptic properties.

Meat Industry

In addition to vegetable preservation, acetic acid is commonly used in meat industry. Widely spread technique employed in meat processing is topical washing with acetic acid solution to reduce bacterial pathogen load by combination of physical removal, temperature and acidity action (Carpenter, Smith & Broadbent, 2011).

In research of Carpenter, Smith & Broadbent (2011), 2% acetic acid prevented residual growth of *E. coli* and *L. monocytogenes*, and reduced number of *Salmonella* on chicken skin below recoverable levels.

Spraying pork cheek meat with 2% acetic acid reduced aerobic plate count and incidence of *Salmonella*, application of acetic acid at 70 °C was effective in sanitizing beef contaminated with *E. coli, S. typhimurium* and manure slurry, and combined action of 2% acetic acid spray and pulsed electricity reduced load of *E. coli* O157:H7 and *S. typhymurium* on beef steaks (Mani-Lopez et al., 2012).

Acidic marination improves techno-functional meat properties: water holding capacity, tenderness, juiciness and colour. Fall in pH and increase in ionic strength change swelling properties and myofibrillar protein extractability. Often, acetic acid is used for marinating. Its effectiveness is linked to hydrogen bonds between acetic acid and protein titratable groups (Goli et al., 2012). However, acetic acid application in meat treatment is limited due to discolouration of meat. Stivarius et al. (2002) studied effect of 5% acetic acid treatment on lean beef trimmings inoculated with *E. coli* and *S. typhimurium*. During 7-day storage acetic acid reduced coliform, aerobic plate counts, *E. coli* and *S. typhimurium*, but meat colour and odour changed as well.

Harris et al. (2012) studied effectiveness of acetic acid (2 and 5%) in reduction of *E. coli* and *S. typhimurium* in simulated processing environment. They reported that acetic acid successfully reduced the number of pathogens, but no significant difference was found between the two concentrations. Initially, acid treatment did not influence meat colour, but after 24 h redness was lower in treated samples.

Dubal et al. (2004) significantly reduced colour change by replacing part of acetic acid with propionic (1.5% acetic + 1.5% propionic) acid and extended shelf life of sheep/goat meat at refrigeration temperature (5 – 7 °C) by double compared to untreated samples. Mixture of buffered vinegar and sodium lactate decreased total plate count during 18-day storage of hot-boned pork sausage patties, and redness and sensory quality (acceptability and juiciness) were maintained for 17 days, unlike untreated samples, which retained acceptable parameters for 14 days (Bradley et al., 2011).

Kakatkar, Sharma & Venugopal (2003) applied 2% acetic acid in gelation of fish protein to obtain product suitable for surimi application. They chose Bombay duck, fish with very high moisture content which makes it unsuitable for surimi production. After holding the fish overnight in cold 2% acetic acid, they observed gelation of fish proteins which wasn't influenced by repeated freezing and thawing.

Active Packaging

"Active packaging is packaging type that changes condition to extend shelf-life or improve safety or sensory properties while maintaining the quality of food" (Quintavalla & Vicini, 2002). The term "active" implies that they interact with the food and the environment – they scavenge oxygen, moisture, etc and/or release antimicrobial substances and aroma compounds, etc (Quintavalla & Vicini, 2002).

Packaging material can be synthetic (e.g. LDPE films) or edible (e.g. chitosan or starch based films) (Ouattara et al., 2000).

If synthetic films are used, antimicrobial agent can be incorporated into the film during production (e.g. extrusion), or layered on the interior surface after the production (Quintavalla & Vicini, 2002). When edible films are produced, most often antimicrobial substances are incorporated into the film (Emiroglu, 2010; Iturriaga, Olabarrieta & Maranon, 2012).

Regardless of film type, antimicrobial substances are slowly released on the food surface, which enables them better activity through prolonged periods of time. In application of active packaging care must be taken in selection of material with proper chemical properties in regard to reactions of packaging with food, environment and additives incorporated into it (Quintavalla & Vicini, 2002).

Acetic acid may be used as a solvent for preparation of chitosan-based edible films (Ouattara et al., 2000; Park et al., 2004; Vasconez, 2009). In addition to chitosan, Jiang et al. (2011) dissolved starch, alginate and pectin in 1% acetic acid to produce edible coatings. Although acetic acid is considered to be solvent, it can act as antimicrobial component of film.

Namely, research of Ouattara et al. (2000) has shown that acetic acid migrates from chitosan film to the meat surface within first 3 h of storage and, according to the same research, film prepared by dissolving chitosan in 1% acetic acid effectively inhibited *Enterobacteriaceae* at meat surface. Suppakul et al. (2003) list acetic acid among antimicrobial components of active packaging and Appendini & Hotchkiss (2002) state that acetic acid is antimicrobial agent in edible films targeting mainly moulds.

Starch Modification

Acetic acid and its anhydride are often used in starch modification. Reaction of these reagents with starch, in which esters are formed, can be catalysed with acids, e.g. 1% H_2SO_4 (Babić, 2007) or alkalies, e.g. NaOH, KOH (Eliasson, 2004). Bulky acetic group sterically hinders H-links between starch chains and therefore starch acetates have lower gelatinisation enthalpies, lower retrogradation, form clearer pastes which are more stable during storage and freeze-thaw cycles (Singh et al., 2007).

Starch acetates are widely used in food production. They are added to bakery products to retard retrogradation and "aging" of the products, in meat products they bound water and reduce loss during cooking, and in miscellaneous products they act as texture improvers and rheology modifiers (BeMiller & Whistler, 2009).

Starch acetates with high degree of substitution are used as tablet binders, biodegradable plastic, as substitute for cellulose acetate, etc (Diop et al., 2011).

Sanitation in Food Industry

Regarding sanitation, acids are generally used for removal of inorganic deposits and descaling. Organic acids are components of ecological detergents, where they are combined with surfactants.

Although it can be used in cleaning of food premises, due to strong odour and low activity, acetic acid is often combined with peracetic acid and hydrogen peroxide. Using cleaners based on acetic acids should be done with care, since it attacks skin and eyes. It also presents higher fire hazard than other commonly used cleaners (Marriot & Gravani, 2006).

However, acetic acid is valuable starting material for production of peracetic acids, which is strong disinfectant (Stanga, 2010), more effective than Cl^- in removal of *Listeria monocytogenes* biofilms (Belessi et al., 2011).

Acetic Acid and Nutritional/Health Aspects

Apple cider vinegar and its products have been considered as aid in high blood-pressure control, weight loss, indigestion, haemorrhoids, etc for many decades (Hill et al., 2005). Brewed vinegar, according to literature data, stimulates appetite, aids digestion, promotes calcium absorption, regulates blood glucose and blood pressure, etc. (Xu, Tao & Ao, 2007). Xu, Tao & Ao (2007) reported that melanoidines formed in Zhenjiang vinegar have antioxidant properties and potential health promoting effect.

Sugiyama et al. (2003) investigated influence of mixed vinegar/wine beverage on cardiovascular function of rats and concluded that this beverage could have beneficial effect on people with palpitation and hypertension. Sediment of Kurosu vinegar (*Kurosu moromimatsu*) is reach in minerals and organic components and inhibits tumor growth of Lovo cells in mice (Fukuyama et al., 2007).

Shishehbor et al. (2008) reported that apple cider vinegar significantly reduced low density lipoprotein-cholesterol in normal rats, and reduced serum triglyceride and increased HDL in diabetic rats. Johnston & Buller (2005) reported that acetic acid reduced postprandial glycaemia after high-glycaemic food consumption due to disaccharidase inhibition in small intestine or by stimulating uptake and utilisation of glucose in peripheral tissues.

However, Hlebowicz et al. (2007) stated that vinegar delays gastric emptying and that it may have adverse effect on glycaemic control in insulin-dependent diabetes mellitus patients and Salbe et al. (2009) reported that "vinegar does not act to decrease glycaemia by interference with enteral carbohydrate absorption".

Remarks for Future Research

Farahnaky & Afshari-Jouybari (2011) investigated potential of hot acetic acid (0.5%; 40 °C) application in accelerated ripening and decay prevention of dates. Their research showed that this process could be used without significant influence on consumer preference. Potential of acetic acid application in other postharvest treatments should also be investigated.

Research of Ohishi et al. (2007) showed that addition of acetic acid during rice cooking has influence on textural properties and stickiness of rice – additional research on exact reaction mechanism of acetic acid with rice and rice starch, and on other starchy food as well, should elucidate this effect.

Additional research is needed on the acetic acid application in meat industry as well. As above mentioned, acetic acid has antimicrobial activity, but also causes discolouration of meat. Although some efforts have been made in elucidating and preventing the latter, undesirable effect, there is still space for progress.

Nutritional and health effect of different vinegar types is also poorly investigated, and there are some controversies which should be resolved.

In conclusion, although vinegar has long been part of human nutrition, it is still a mystery to us in great deal. Therefore, it could be very interesting material for research both in food technology and nutrition.

References

Andres-Barrao, C; Saad, MM; Chappuis, ML; Boffa, M; Perret, X; Perrez, RO; Barja, F. Proteome analysis of *Acetobacter pateurianus* during acetic acid fermentation. *Journal of Proteomics,* 2012 75, 1701-1717.

Appendini, P; Hotchkiss JH. Review of antimicrobial food packaging. *Innovative Food Science and Emerging Technologies,* 2002 3, 113-126.

Babić, J; Šubarić, D; Ačkar, Đ; Kovačević, D; Piližota, V; Kopjar, M. Preparation and characterization of acetylated tapioca starches. *Deutsche Lebensmittel-Rundschau,* 2007 103, 580-585.

Belessi, CEA; Gounadaki, AS; Psomas, AN; Skandamis, PN. Eficiency of different sanitation methods on *Lysteria monocytogenes* biofilms formed under various environmental conditions. *International Journal of Food Microbiology,* 2011 145, S45-S52.

BeMiller JN; Whistler, RL. *Starch chemistry and technology.* 3rd ed. USA: Academic Press; 2009.

Blech, ZY. *Kosher food production.* 2nd Ed. USA: Wiley & Blackwell; 2009.

Bradley, EM; Williams, JB; Schilling, MW; Coggins, PC; Crist, C; Yoder, S; Campano, SG. Effects of sodium lactate and acetic acid derivatives on the quality and sensory characteristics of hot-boned pork sausage patties. *Meat Science,* 2011 88, 145-150.

Carpenter, CE; Smith JV; Broadbent, JR. Efficacy of washing meat surfaces with 2% levulinic, acetic, or lactic acid for pathogen decontamination and residual growth inhibition. *Meat Science,* 2011 88, 256-260.

Chen, C; Chen, F. Study on the conditions to brew rice vinegar with high content of γ-amino butyric acid by response surface methodology. *Food and Bioproducts Processing,* 2009 87, 334-340.

Croatian legislation: Pravilnik o vinskom i voćnom octu uz izmjene i dopune (NN 121/05; 53/06; 26/11).

Diop, CIK; Li, HL; Xie, BJ; Shi, J. Effects of acetic acid/acetic anhydride ratios on the properties of corn starch acetates. *Food Chemistry,* 2011 126, 1662-1669.

Dubal, ZB; Paturkar, AM; Waskar, VS; Zende, RJ; Latha, C; Rawool, DB; Kadam, MM. Effect of food grade organic acids on inoculated *S. aureus, L. monocytogenes, E. coli* and *S. typhimurium* in sheep/goat meat stored at refrigeration temperature. *Meat Science,* 2004 66, 817-821.

Eliasson, AC. *Starch In Food.* Cambridge, England: Woodhead Publishing; 2004.

EUFIC Food Additives Approved by the EU. Available from: *www.eufic.org/upl/1/.../e-numbers_eufic(2).pdf* (Accessed: 11/03/2013)

Emiroglu, ZK; Yemis, GP; Coskun, BK; Candogan, K. Antimicrobial activity of soy edible films incorporated with thyme and oregano essential oils on fresh ground beef patties. *Meat Science,* 2010 86, 283-288.

FAO/WHO Food Standards programme. *Proposed Draft Revised Regional Standard for Vinegar.* Rome, Italy; 2000.

Farahnaky, A; Afshari-Jouybari, H. Physicochemical changes in Mazafati dates fruits incubated in hot acetic acid for accelerated ripening to prevent diseases and decay. *Scientia Horticulturae,* 2011 127, 313-317.

Fernandez, AR; Mira, NP; Vargas, RC; Canelhas, I; Sa-Correia, I. *Saccharomyces cerevisiae* adaptation to weak acids involves the transcription factor Haa1p Haa1p-regulated genes. *Biochemical and Biophysical Research Communications,* 2005 337, 95-103.

Fukuyama, N; Jujo, S; Ito, I; Shizuma, T; Myojin, K; Ishiwata, K; Nagano, M; Nakazawa, H; Mori, H. Kurozu moromimatsu inhibits tumor growth of Lovo cells in a mouse model in vivo. *Nutrition,* 2007 23, 81-86.

George, SM; Richardson, LCC; Peck, MW. Predictive models of the effect of temperature, pH and acetic and lactic acids on the growth of *Lysteria monocytogenes. International Journal of Food Microbiology,* 1996 32, 73-90.

Goli, T; Bohuon, P; Ricci, J; Collignan, A. Evolution of pH during immersion of meat protein matrices in acidic marinades. *Meat Science,* 2012 90, 618-623.

Greenacre, EJ; Brocklehurst, TF. The acetic acid tolerance response induces cross-protection to salt stress in *Salmonella typhimurium. International Journal of Food Microbiology,* 2006 112, 62-65.

Harris, D; Brashears, MM; Garmyn, AJ; Brooks, JC; Miller, MF. Microbiological and organoleptic characteristics of beef trim and ground beef treated with acetic acid, lactic acid, acidified sodium chlorite, or sterile water in a simulated commercial processing environment to reduce *Escherichia coli* O157:H7 and *Salmonella. Meat Science,* 2012 90, 783-788.

Hill, LL; Woodruff, LH; Foote, JC; Barreto-Alcoba, M. Esophageal injury by apple cider vinegar tablets and subsequent evaluation of products. *Journal of the American Dietetic Association,* 2005 105, 1141-1144.

Hlebowicz, J; Darwiche, G; Bjorgell, O; Almer, LO. Effect of apple cider vinegar on delayed gastric emptying in patients with type I diabetes mellitus: a pilot study. *BMC Gastroenterology,* 2007 7, 46-51.

Iturriaga, L; Olabarrieta, I; de Maranon, IM. Antimicrobial assays of natural extracts and their inhibitory effect against *Lysteria innocua* and fish spoilage bacteria, after incorporation into biopolymer edible films. *International Journal of Food Microbiology,* 2012 158, 58-64.

Jiang, Z; Neetoo, H; Chen, H. Efficacy of freezing, frozen storage and edible antimicrobial coatings used in combination for control of *Listeria monocytogenes* on roasted turkey stored at chiller temperatures. *Food Microbiology,* 2011 28, 1394-1401.

Johnston, CS; Buller, AJ. Vinegar and peanut products as complementary foods to reduce postprandial glycemia. *Journal of the American Dietetic Association,* 2005 105, 1939-1942.

Kakatkar, A; Sharma, A; Venugopal, V. Hydration of muscle proteins of Bombay duck (*Harpodon nehereus*) during acetic acid-induced gelation and characteristics of the gel dispersion. *Food Chemistry,* 2003 83, 99-106.

Lee, SY; Kang, DH. Combined effects of heat, acetic acid, and salt for inactivating *Escherichia coli* O157:H7 in laboratory media. *Food Control,* 2009 20, 1006-1012.

Leistner, L. Basic aspects of food preservation by hurdle technology. *International Journal of Food Microbiology,* 2000 55, 181-186.

Leon Pelaez, AM; Catano, CAS; Yepes, EAQ; Villarroel, RRG; De Antoni GL; Giannuzzi, L. Inhibitory activity of lactic and acetic acid on *Aspergillus flavus* growth for food preservation. *Food Control,* 2012 24, 177-183.

Lovrić, T; Piližota, V. *Konzerviranje i prerada voća i povrća.* Zagreb, Croatia: Nakladni zavod Globus; 1994.

Luck, E; Jager, M. *Antimicrobial food additives: characteristics, uses, effects.* 2nd Ed. Berlin: Springer-Verlag; 1997.

Mani Lopez, E; Garcia, HS; Lopez-Malo, A. Organic acids as antimicrobials to control *Salmonella* in meat and poultry products. *Food Research International,* 2012 45, 713-721.

Marriot, NG; Gravani, RB. *Principles of food sanitation.* 5th Ed. USA: Springer; 2006.

Mert, B. Using high pressure microfluidization to improve physical properties and lycopene content of ketchup type products. *Journal of Food Engineering,* 2012 109, 579-587.

Ohishi, K; Kasai, M; Shimada, A; Hatae, K. Effects of acetic acid on the rice gelatinisation and pasting properties of rice starch during cooking. *Food Research International,* 2007 40, 224-231.

Ouattara, B; Simard, RE; Piette, G; Begin, A; Holley, RA. Inhibition of surface spoilage bacteria in processed meats by application of antimicrobial films prepared with chitosan. *International Journal of Food Microbiology,* 2000 62, 139-148.

Park, SI; Daeschel, MA; Zhao, Y. Functional properties of antimicrobial lysozyme-chitosan composite films. *Journal of Food Science,* 2004 69, M215-M221.

Quintavalla, S; Vicini, L. Antimicrobial food packaging in meat industry. *Meat Science,* 2002 62, 373-380.

Ranken, MD; Baker, CGJ, Kill, RC. *Food Industries Manual.* 24th Ed. USA: Chapmann & Hall, 1997.

Retief, F; Cilliers, LP. Lead poisoning in Ancient Rome. *Acta Theologica,* 2006 26, 147-164.

Sahin, H; Ozdemir, F. Effect of some hydrocolloids on the rheological properties of different formulated ketchups. *Food Hydrocolloids,* 2004 18, 1015-1022.

Salbe, AD; Johnston, CS; Buyukbese, MA; Tsitouras, PD; Harman, SM. Vinegar lacks antiglycemic action on enteral carbohydrate absorption in human subjects. *Nutrition Research,* 2009 29, 846-849.

Schleputz, T; Gerhards, JP; Buchs, J. Ensuring constant oxygen supply during inoculation is essential to obtain reproducible results with obligatory aerobic acetic acid bacteria in vinegar production. *Process Biochemistry,* 2013 In Press, doi:10.1016/ j.procbio.2013.01.009

Sengun, IY; Karabiyikli, S. Importance of acetic acid bacteria in food industry. *Food Control,* 2011 22, 647-656.

Shishehbor, F; Mansoori, A; Sarkaki, AR; Jalali, MT; Latifi, SM. Apple cider vinegar attenuates lipid profile in normal and diabetic rats. *Pakistan Journal of Biological Sciences,* 2008 11, 2634-2638.

Singh, J; Kaur, L; McCarthy, OJ. Factors influencing the physico-chemical, morphological, thermal and rheological properties of some chemically modified starches for food applications - a review. *Food Hydrocolloids* 2007 21, 1-22.

Solieri, L; Giudici, P. *Vinegars of the World.* Italia: Springer-Verlag; 2009.

Stanga, M. *Sanitation – Cleaning and Disinfection in Food Industry.* Germany: Wiley-Verlag; 2010.

Stivarius, MR; Pohlman, FW; McElyea, KS; Apple, JK. The effects of acetic acid, gluconic acid and trisodium citrate treatment of beef trimmings on microbiological, color and odor characteristics of ground beef through simulated retail display. *Meat Science,* 2002 60, 245-252.

Stratford, M; Plumridge, A; Nebe-von-Caron, G; Archer, DB. Inhibition of spoilage mould conidia by acetic acid and sorbic acid involves different modes of action, requiring modificatrion of weak-acid theory. *International Journal of Food Microbiology,* 2009 136, 37-43.

Sugiyama, A; Saitoh, M; Takahara, A; Satoh, Y; Hashimoto, K. Acute cardiovascular effects of a new beverage made of wine vinegar and grape juice, assessed using an in vivo rat. *Nutrition Research,* 2003 23, 1291-1296.

Suppakul, P; Miltz, J; Sonneveld, K; Bigger, SW. Active packaging technologies with an emphasis on antimicrobial packaging and its applications. *Journal of Food Science,* 2003 68, 408-420.

Tesfaye, W; Morales, ML; Garcia-Parilla, MC; Troncoso, AM. Wine vinegar: technology, authenticity and quality evaluation. *Trends in Food Science and Technology,* 2002 13, 12-21.

Theron, MM; Lues, JFR. *Organic Acids and Food Preservation.* USA: CRC Press; 2011.

Vasconez, MB; Flores, SK; Campos, CA; Alvarado, J; Gerschenson, LN. Antimicrobial activity and physical properties of chitosan-tapioca starch based edible films and coatings. *Food Research International,* 2009 42, 762-769.

Vermeulen, A; Gysemans, KPM; Bernaerts, K; Geeraerd, AH; Van Impe, JF; Debevere, J; Devlieghere F. Influence of pH, water activity and acetic acid concentration on *Listeria monocytogenes* at 7°C: data collection for the development of a growth/no growth model. *International Journal of Food Microbiology,* 2007 114, 332-341.

Wu, JJ; Ma YK; Zhang, FF; Chen, FS. Biodiversity of yeasts, lactic acid bacteria and acetic acid bacteria in the fermentation of "Shanxi aged vinegar", a traditional Chinese vinegar. *Food Microbiology,* 2012 30, 289-297.

Xu, W; Huang, Z; Zhang, X; Li, Q; Lu, Z; Shi, J; Xu, Z; Ma, Y. Monitoring the microbial community during solid state acetic acid fermentation of Zhenjiang aromatic vinegar. *Food Microbiology,* 2011 28, 1175-1181.

Xu, Q; Tao, W; Ao, Z. Antioxidant activity of vinegar melanoidins. *Food Chemistry,* 2007 102, 841-849.

In: Acetic Acids
Editor: Angelo Basile
ISBN: 978-1-62948-217-0

Chapter 8

Coproduction of Vinegar and Electricity by Application of Microbial Fuel Cell Technology to Acetic Acid Fermentation

Takanori Tanino[*]
Division of Environmental Engineering Science, Faculty of Science and Technology, Gunma University, Kiryu, Gunma, Japan

Abstract

It is well known that vinegar is one of the most familiar uses of acetic acid. In this chapter, the coproduction of vinegar and electricity by fermentation, which is a novel concept in vinegar production, is explained. The acetic acid in vinegar is produced from ethanol by acetic acid fermentation using acetic acid bacteria. Acetic acid fermentation is a typical example of a traditional application of microorganisms. In addition to traditional uses of microorganisms, the fields of microorganism application have been expanding owing to the progress of biotechnology. Among these fields, microbial fuel cell technology has recently been attracting attention as a novel method for harvesting energy from renewable organic compounds. The novel concept of the coproduction of vinegar and electricity has been developed by the application of microbial fuel cell technology to acetic acid fermentation. Acetic acid fermentation and microbial fuel cell technology share the same principle in that microorganisms are used as the catalyst for the oxidation of an organic compound. In this chapter, the core elements of acetic acid fermentation and microbial fuel cell technology that are important for understanding our novel concept are first described. Then, the advantages and our recent study of the coproduction of acetic acid and electricity are described.

[*] Corresponding author: Takanori Tanino. Division of Environmental Engineering Science, Faculty of Science and Technology, Gunma University, 1-5-1, Tenjincho, Kiryu, Gunma 376-8515, Japan. E-mail: ttanino@gunma-u.ac.jp.

Introduction

This chapter gives an outline of the coproduction of acetic acid and electricity by fermentation. This novel concept has recently been demonstrated in our laboratory and was achieved by applying microbial fuel cell technology to acetic acid fermentation [1]. The ultimate objective of our research is the coproduction of commercially available vinegar and electricity. It is well known that acetic acid is the main component of vinegar, which is one of the most familiar uses of the acetic acid. The raw materials of vinegar are ethanol solutions produced by alcoholic fermentation catalyzed by microorganisms from juices of various crops that contain alcoholic fermentable sugars. Therefore, there are various types of vinegars around the world in which the raw materials differ, all of which are produced by fermentation catalyzed by acetic acid bacteria, i.e., acetic acid fermentation. In Europe and the United States, white vinegar, which is fermented from distilled ethanol and used as a condiment and in pickle production, is familiar and accounts for the largest amount of vinegar production. In Asia, in addition to white vinegar for food processing, a similar amount of cereal vinegar is also produced and used as a condiment. Vinegar production has a long history, and several dedicated reactors for large-scale commercial production, such as acetators and cavitators, had been developed by the mid -twentieth century. Today's vinegar production processes, especially those for white vinegar and cereal vinegar, which are the most suitable and largest-scale processes to which our novel conceptcan be applied, are well-established commercial processes using traditional, biological and engineering techniques and knowledge. Our novel concept has only recently been verified at the laboratory scale; therefore, it cannot compete with today's vinegar production processes without considerable further research. However, as described below, our novel concept has several advantages including energy production, and it has the potential to be an important technique for future vinegar production processes. Before giving a detailed description of our novel concept, the core elements of acetic acid fermentation and microbial fuel cell technology that are important for understanding our novel concept are described below.

Acetic Acid Fermentation

Acetic acid fermentation is a typical example of oxidation fermentation, in which bacteria oxidize substrates while consuming oxygen and large amounts of metabolic intermediates are accumulated in the culture media.

As well as alcoholic fermentation, acetic acid fermentation is a process that takes advantage of microbial metabolism to produce useful materials. The bacteria that carry out acetic acid fermentation are called acetic acid bacteria and are categorized in the *Acetobacteraceae* family and as gram-negative aerobes [2].

Among the acetic acid bacteria, bacteria categorized as *Acetobacter* species and *Gluconobacter* species play important roles in vinegar production. As well as yeast and other microorganisms, which have been used to produce foods for a long time, it is considered that acetic acid bacteria are generally recognized as safe (GRAS). Two enzymes of acetic acid bacteria catalyze acetic acid fermentation, and these enzymes are localized on the surface of the acetic acid bacteria cell membrane (Figure 1).

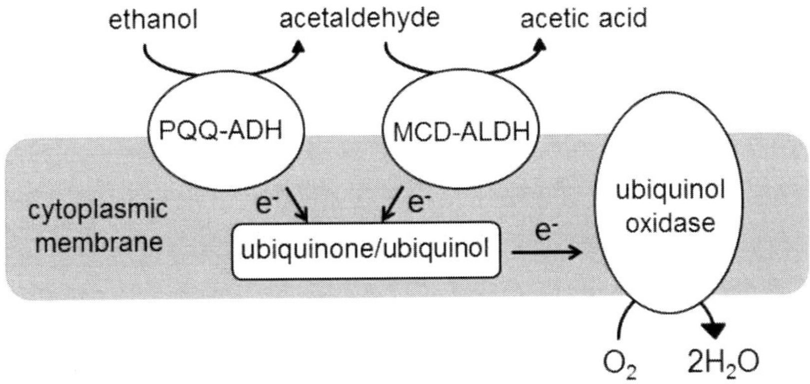

Figure 1. Fermentation process on the acetic acid bacteria cell surface.

The oxidization of ethanol to acetaldehyde is catalyzed by alcohol dehydrogenase (ADH), which contains the specific coenzyme pyrroloquinoline quinone (PQQ) and heme c [3]. Redox enzymes such as ADH, which contains PQQ as the coenzyme, are called quinoproteins. The oxidation of acetaldehyde to acetic acid is catalyzed by aldehyde dehydrogenase (ALDH), which contains the coenzyme that is considered to be Molybdopterin cytosine dinucleotide (MCD) and heme c. This mechanism of acetic acid fermentation catalyzed by these two enzymes localizing on the cytoplasmic membrane and containing PQQ, which was mainly revealed by the respected and intensive research of agroup at Yamaguchi University (Matsushita, Adachi and Ameyama) [4-6], is a very distinctive phenomenon. The oxidation reactions of ethanol and acetoaldehyde are directly connected to the respiratory chain on the surface of the cell membrane. The oxidation of one ethanol molecule to acetic acid generates four electrons. The electrons generated from the oxidation reactions are used to reduce ubiquinone, which is localized in the cell membrane, to ubiquinol. Ubiquinol is oxidized by ubiquinol oxidase localized at the cell membrane, and electrons are donated to oxygen molecules.

Microbial Fuel Cell Technology

It is well known that microorganisms acquire energy through the metabolism of organic compounds and that their own enzymes catalyze metabolic reactions. Microbial fuel cell technology, which enables the direct conversion of the chemical energy contained in organic compounds into electrical energy by microbial action, has been attracting attention as a novel method for harvesting energy from renewable biomass resources. In the last decade, microbial fuel cell technology has been one of the most active research areas, and a lot of research papers have been published. There have also been numerous review papers, in which recent developments and applications of microbial fuel cell technology are well summarized [7-12]. Therefore, only an outline of microbial fuel cell technology, which is required to understand our novel concept is described below.

Microbial fuel cell technology is not constrained by Carnot's cycle, and high energy conversion efficiency can be achieved in theory. However, in reality there are several factors that reduce the energy conversion efficiency.

Under usual aerobic conditions, the oxidation reactions of the organic compound, in which electrons and protons are released, and the reduction reaction of oxygen, in which the released electrons and protons are consumed, proceed in the same space. In microbial fuel cell technology, the oxidation reactions of the organic compound are carried out under anaerobic conditions to produce electrons that are donated to the electrode, similarly to in other fuel cell technologies, and redox reactions are carried out in a spatially partitioned space (Figure 2). The electrode that accepts electrons produced from the oxidation reaction of the organic compound catalyzed by microorganisms is called the anode, and the electrode that donates electrons to the electron acceptor is called the cathode. To separate the redox reactions, a proton/cation exchange membrane is used. A Nafion (DuPont Co., Wilminton, US) membrane is the most well-known and widely used membrane in microbial fuel cell technology, but unfortunately Nafion membranes are relatively expensive. There are two typical microbial fuel cell structures; two-chamber and single-chamber structures. In the two-chamber structure, the cathode is placed in catholyte. On the other hand, in the single-chamber structure, the cathode is exposed directly to air. The anode chamber in which the oxidation of the organic compound by the microorganism occurs is kept under anaerobic conditions, and in many cases chemicals that can become involved in the oxidation reaction and remove electrons from the microbial cell, which are called electron mediators, are added to the anode reaction mixture. Dyes (e.g. thionine, methylene blue, neutral red) and vitamins (e.g., vitamin K_3, vitamin B_{12}) are often used as electron mediators, and electrons accepted by these chemicals and removed from the microbial cell are donated to the anode. Because of the potential difference between the anode and cathode, the electrons move through the external circuit to the cathode, on which the reduction reaction is carried out. The protons released from the oxidation of the organic compound also move to the anode through the separating membrane by diffusion. The electrons and protons react with the electron acceptor on the cathode. Oxygen or ferricyanide is often used as the electron acceptor. Carbon materials are usually used for both electrodes. Platinum has been used as the catalyst for the cathode but platinum is very expensive.

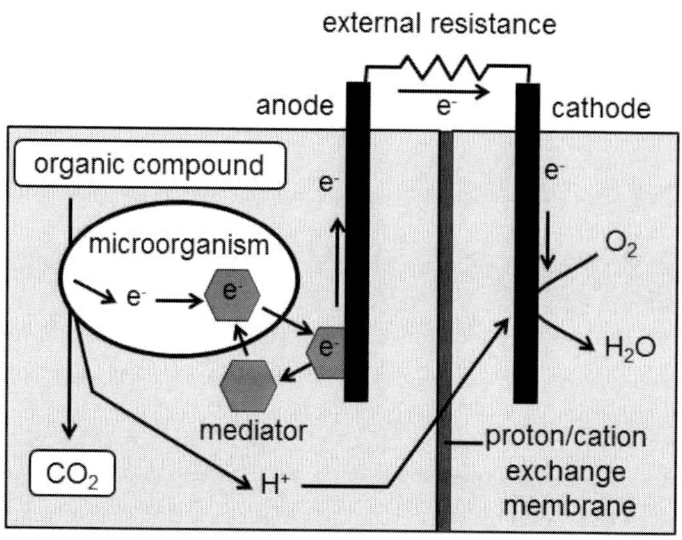

Figure 2. Microbial fuel cell structure.

Recently, to reduce the cost of microbial fuel cells, the use of an alternative catalyst for the cathode such as a carbon alloy or a microorganisms has been investigated. Renewable biomass is an attractive source of energy because of its carbon-neutral characteristics. The most active fields of application of microbial fuel cell technology are wastewater treatment processes and mobile power supplies. In the application to wastewater treatment processes, organic waste is a suitable energy resource, but to accomplish the goal of cleaning water and reducing the amount of sludge as part of a wastewater treatment protocol, organic waste must be oxidized into CO_2 via microbial metabolism. In the application to mobile power supplies, a high-energy-density material such as a sugar or alcohol is adopted as the energy resource, which must also be oxidized into CO_2 via microbial metabolism to realize a compact power supply.

Advantages of Coproduction of Acetic Acid and Electricity

As mentioned above, acetic acid fermentation and microbial fuel cell technology share the same principle that microorganisms are used to catalyze the oxidation of an organic compound. We anticipate that the application of microbial fuel cell technology for vinegar production will enable the simultaneous production of electricity and acetic acid.

Bioelectrocatalysis using quinoproteins from several bacteria including acetic acid bacteria has been used in biosensors for several sugars and alcohols and in biofuel cells [13]. Moreover, the characteristics of quinoproteins from *Gluconobacter* species have been well investigated [14]. In these studies, it was revealed that not only purified enzymes but also intact whole bacteria cells can carry out bioelectrocatalysis. Recently, it has been reported that the acetic acid bacteria *Acetobacter aceti* and *Gluconobacter roseus* can produce electricity from glucose and ethanol in microbial fuel cell without the addition of electron mediators [15, 16]. The enzyme that catalyzes glucose oxidation is glucose dehydrogenase (GDH), which is also a quinoprotein and is localized on the outer membrane of acetic acid bacteria. The mechanism by which acetic acid bacteria can produce electricity from glucose and ethanol in a microbial fuel cell without the addition of electron mediators must be closely related to the fact that the oxidation reactions of glucose and ethanol are catalyzed by quinoproteins at the cell surface of acetic acid bacteria. In the case of using ethanol as the energy source, the characteristic of acetic acid bacteria, that no electron mediators are necessary to transfer electrons from acetic acid bacteria to the electrode makes it favorable for ensuring the quality of vinegar produced in a microbial fuel cell.

In conventional acetic acid fermentation under aerobic conditions, the energy generated from the oxidation of ethanol is eventually converted to fermentation heat, increasing the temperature of the fermentation culture. In acetic acid fermentation using dedicated reactors for large-scale commercial production, a large amount of fermentation heat is generated, and an excessive increase in the temperature of the fermentation culture inactivates the acetic acid bacteria. Therefore, a cooling system, which uses energy, is required for the reactor. In the coproduction of acetic acid and electricity by applying microbial fuel cell technology to acetic acid fermentation, it is expected that most of the energy generated from the oxidation of ethanol will be converted to electric energy.

Therefore, the generation of fermentation heat and the increase in temperature of the fermentation culture caused by fermentation heat can be avoided. Moreover, the amount of energy required to cool the reactor is reduced. Energy conservation is thus possible in vinegar production by acetic acid fermentation by reducing the consumption of energy for cooling while generating electric energy. We anticipate that this advantage will create novel value and lead to innovation in vinegar production.

In the application of microbial fuel cell technology, the coproduction of acetic acid and electricity by applying microbial fuel cell technology to acetic acid fermentation also has several different points from conventional research. A comparison of our new concept and conventional researche in the application of microbial fuel cell technology is given in Table 1. As mentioned in the section on microbial fuel cell technology, the most active fields of application of microbial fuel cell technology are wastewater treatment processes and mobile power supplies.

In both application fields, an organic compound must be oxidized to CO_2 (complete oxidation) via microbial metabolism to accomplish the objective. The accumulation of metabolic intermediates resulting from the incomplete oxidation of organic compounds is a problem to solve.

Therefore, no studies in which a desirable metabolic intermediate has been accumulated have been reported. In our novel concept, the accumulation of a metabolic intermediate, acetic acid, by incomplete oxidation is the objective, and complete oxidation must be prevented.

All the energy sources in these applications are renewable and carbon-neutral biomass resources. In the application to wastewater treatment, a diverse range of organic compounds can be adopted as the energy source because of many types of wastewater.

Indeed, studies on treating various types of wastewater such as domestic wastewater [17, 18], dye wastewater [19, 20], brewery wastewater [21, 22] and starch- and potato-processing wastewater [23, 24] have been reported. In the application to mobile power supplies, high-energy-density energy sources such as sugars and alcohols are suitable to reduce the cell size. In the application to vinegar production, the energy source is edible ethanol used for vinegar production. It is produced by fermentation catalyzed by microorganisms from the juices of various crops that contain alcoholic fermentable sugars and, if necessary, by the distillation of alcoholic fermentation products.

Table 1. Comparison of microbial fuel cell applications

	Previous studies		Our novel concept
Application	Wastewater treatment	Mobile power supply	Vinegar production
Oxidation	Complete	Complete	Incomplete
Energy source	Diverse kinds of organic compound in wastewater	High-energy density organic compound (sugar or alcohol)	Edible ethanol produced by fermentation from juices of various crops
Microorganism	Various species employed in wastewater treatment	Ideal microorganisms including genetic modified one	Acetic acid bacteria (GRAS)
System size	Large	Small	Medium

However, the electric energy recovered by microbial fuel cell technology in this application is the energy that is changed into fermentation heat, resulting in the consumption of energy for cooling in conventional vinegar production; this energy has not been previously used as an energy source. Therefore, the use of edible ethanol in this application does not compete with the food problem.

The characteristics of microorganisms that can be used in these applications also have several differences. In the application to wastewater treatment, microorganisms contained in the anode chamber are usually a mixture of various species generally employed in wastewater treatment. In the application to mobile power supplies, ideal microorganisms for producing electricity from organic compounds, such as *Shewanella* species and *Geobacter* species that have electrically conductive bacterial nanowires, are adoptable and, in the future, genetically modified bacteria may be used. In the application to vinegar production, the only microorganisms that can be used in the anode chamber are acetic acid bacteria, which are GRAS and for which there is a large amount of available information based on traditional and biological backgrounds.

In microbial fuel cell technology, the voltage that can be generated by a single cell is low, similarly to in other fuel cell technologies. Therefore, the construction of a system by assembling single-cell structures to generate a sufficient voltage is required. However, the maximum sizes of a single cell and system are directly determined by the type of microbial fuel cell application itself. The amount of wastewater differs with the emission source, but the quantities of both industrial and domestic wastewater produced are generally abundant. To treat such wastewater satisfactorily, the size of single cells can be relatively large, and the size of the system comprising single microbial fuel cells can also be large. In contrast, both the single cell and the system in the case of microbial fuel cells for mobile power supplies must be small and compact. In the application to vinegar production, the use of small single cells, similarly to those used for mobile power supplies, to construct a system would not be advisable from the viewpoints of operation and maintenance in efficient commercial vinegar production. This is because each single cell should produce a significant quantity of vinegar. Therefore, we assume the use of medium-size single cells and that the system comprising single cells will be slightly larger than the size of conventional dedicated reactors used for vinegar production.

Coproduction of Acetic Acid and Electricity

Recently, we have successfully demonstrated the coproduction of acetic acid and electricity by applying microbial fuel cell technology to acetic acid fermentation [1]. Although we performed a small-scale experiment using an experimental fermentation medium in the laboratory, the production of electricity during the production of acetic acid was confirmed. To the best of our knowledge, this is the first report of the application of microbial fuel cell technology to the coproduction of a useful material and electricity. In addition, the coproduction of vinegar using edible ethanol in the microbial fuel cell was also successful. An outline of our study is given below.

The microbial fuel cell structure and the materials we used in our previous study are described first.

We adopted an air-cathode single-chamber microbial fuel cell structure (Figure 3). In this structure, the cathode is directly exposed to air and no catholyte containing specific chemicals is required. Therefore, there is no risk of contamination of the product in the anode chamber by the leakage of chemicals through the membrane. Similarly to in many other studies, a carbon cloth was used as the anode material. The cloth was coated with carbon black containing Nafion as a microporous layer to increase its surface area. Unlike metal materials, such carbon materials and Nafion do not react with acetic acid or elute undesirable ions. Therefore, there is also no risk of contamination of the product in the anode chamber with eluted materials from the electrode or reactants produced by reactions between acetic acid and electrode materials. As the proton/cation exchange membrane, we used a Neosepta membrane (Astom Co., Tokyo, Japan). This membrane is inexpensive, has high stability toward chemicals and is used in several food production processes. These characteristics of the above structure and the materials used in the microbial fuel cell make them suitable for vinegar production in a microbial fuel cell. Moreover, ensuring the safety of materials used in food and electricity coproduction by microbial fuel cell technology is important.

As the acetic acid bacteria, we used the *Acetobacter aceti* NBRC14814 strain. *A. aceti* is one of the most well-known strains of *Acetobacter* species. *A. aceti* was cultivated in a liquid experimental medium before it was used in the anode chamber of the air-cathode single-chamber microbial fuel cell. The cultivated *A. aceti* was collected and resuspended in the experimental fermentation medium.

The coproduction of acetic acid and electricity using the air-cathode single-chamber microbial fuel cell and *A. aceti* suspension was examined. The anode chamber of the microbial fuel cell was filled with the *A. aceti* suspension, and the inlet was immediately sealed to create an anaerobic environment within the anode chamber. The microbial fuel cell was incubated at 30°C, and the suspension in the anode chamber was gently stirred by magnetic stirring. The open-circuit voltage (OCV), which is the voltage between the anode and cathode measured in the absence of current, of the microbial fuel cell was measured immediately after filling the anode chamber with the *A. aceti* suspension.

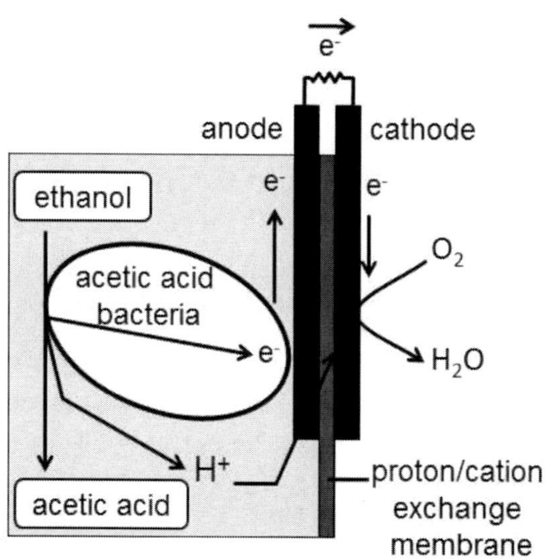

Figure 3. Air-cathode single-chamber microbial fuel cell structure.

The OCV of the microbial fuel cell increased with time and reached a plateau at 0.340 V. The maximum current and power densities after the OCV of the microbial fuel cell reached a plateau were 6.98 μA/cm^2 and 0.818 μW/cm^2 (at 3.41 μA/cm^2), respectively. From these results, it was confirmed that the microbial fuel cell used this study can operate as a fuel cell.

It was examined whether acetic acid fermentation, i.e., the accumulation of acetic acid, can occur in the anode chamber of the microbial fuel cell by connecting the anode and cathode by an external resistance, which allows electrons to move from the anode to the cathode. It was also examined whether the voltage of the microbial fuel cell is maintained when an external resistance connects the anode and cathode. Although the rate of acetic acid production was low (1.23 mmol/L/day), acetic acid production associated with the consumption of ethanol was detected. This showed that acetic acid fermentation, which is the incomplete oxidation of ethanol, is possible in the anode chamber of the microbial fuel cell. The voltage of the microbial fuel cell was also low (approximately 20 mV), but this voltage was maintained during the fermentation of acetic acid. From these results, it was confirmed that the coproduction of acetic acid and electricity by the application of microbial fuel cell technology is possible. The acetic acid yield resulting from the consumption of ethanol during the coproduction of acetic acid and electricity was 73.1%. This is almost equal to that obtained by aerobic fermentation carried out using the same *A. aceti* strain in our laboratory (75.8%). Therefore, it was revealed that the coproduction of electricity did not affect the acetic acid yield resulting from the consumption of ethanol.

After the batch coproduction of acetic acid and electricity, the performances of the microbial fuel cell were higher than those before the coproduction. The OCV, maximum current density and power density were 0.420 V, 16.3 μA/cm^2 and 1.87 μW/cm^2 (at 6.67 μA/cm^2), respectively. Also, the formation of a small amount of biofilm on the anode surface was observed. In research on microbial fuel cells, the immobilization of microorganisms on the anode surface conduced artificially or by promoting biofilm production is thought to be one of the important factors improving the performance of microbial fuel cells. To confirm the effect of biofilm formation on the anode surface on the coproduction of acetic acid and electricity, repeated batch coproduction was carried out without cleaning the anode surface of the microbial fuel cell after the first batch coproduction.

After the second batch coproduction, the formation of a biofilm on the anode surface was confirmed. The OCV increased with the total coproduction time and reached the largest value of 0.521 V prior to the third batch coproduction. However, the OCV decreased after the third batch coproduction, which might have resulted from the inactivation of *A. aceti* during the long operation time.

In the second and third batch coproductions, the maximum current densities and power densities before coproduction, which was after filling the anode chamber with a fresh medium, were smaller than those measured after the previous batches. In both batch coproductions, unsteady voltage profiles were obtained in the initial period of coproduction, indicating that *A. aceti* required time to adapt to the new coproduction conditions. The largest maximum current and power densities of the microbial fuel cell in this study were 19.1 μA/cm^2 and 2.47 μW/cm^2, respectively, after the second batch coproduction. The comparison of *i-V* relationships of the microbial fuel cell before the first batch coproduction and after the second batch coproduction clearly shows the improved performance of the microbial fuel cell (Figure 4).

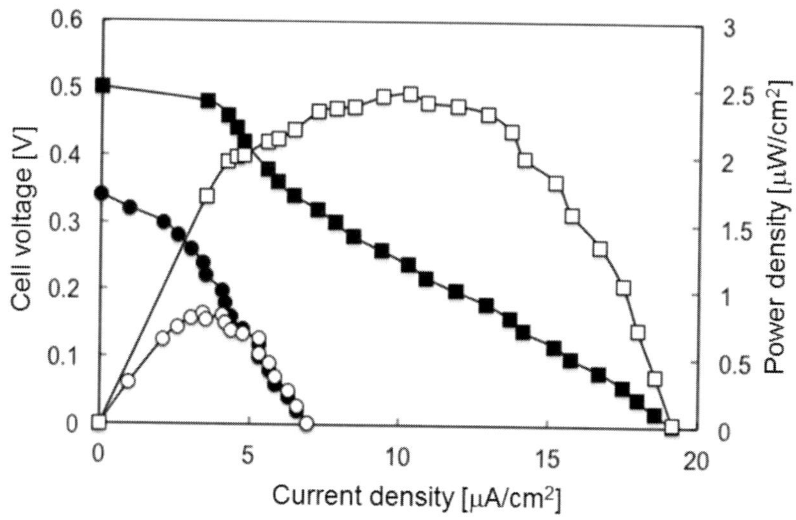

Figure 4. Comparison of cell performance; symbols show cell voltage before the first batch coproduction (closed circles) and after second batch coproduction (closed squares) and power densities before the first batch coproduction (open circles) and after second batch coproduction (closed squares).

From these results, it was demonstrated that the formation of a biofilm on the anode surface improves the performance of the microbial fuel cell. In the third batch coproduction, an increase in the acetic acid productivity (3.36 mmol/L/day) was observed, but no change in the acetic acid yield resulting from the consumption of ethanol was observed (73.2%). Therefore, it was confirmed that the formation of a biofilm on the surface of the anode is important for improving the performance of the coproduction of acetic acid and electricity.

We also demonstrated the coproduction of vinegar and electricity using an edible alcohol. We used *sake*, which is a traditional Japanese alcohol made from rice. *A. aceti* was suspended into diluted *sake*, and the suspension was injected into the microbial fuel cell. The coproduction of vinegar and electricity was carried out under the same conditions as those using the experimental fermentation medium.

Although the OCV, maximum current density and power density of the microbial fuel cell were smaller than those in the case of using the experimental fermentation medium, the voltage of the microbial fuel cell was maintained during the fermentation of acetic acid by connecting the anode and the cathode by an external resistance. The fermentation product contained acetic acid and its taste was slightly sour.

Conclusion

In this chapter, our novel concept, the coproduction of vinegar and electricity by application to acetic acid fermentation, was explained. Although the performance of coproduction is still low, we consider that improving the microbial fuel cell structure, such as by the construction of a suitable reactor architecture, the selection of electrode materials and modification of the electrode surface, will improve the performance of the microbial fuel cell. However, it is important to remember that the safety of these improvements of the structure and materials should be confirmed to ensure safety in food production.

To produce a voltage and commercially available vinegar, the construction of the system from single cells must also be improved. It was confirmed that the coproduction of electricity did not affect the acetic acid yield resulting from the consumption of ethanol. However, vinegar contains several other organic acids, which are important for its quality and taste. A detailed investigation of the effet of coproduction on the composition of these organic acids in vinegar produced by the coproduction of vinegar and electricity is required. Since we have only recently started research on our novel concept, there are many problems to be solved. However, we believe that our novel concept will be an important technique in future vinegar production processes.

Acronyms

ADH: alcohol dehydrogenase
PQQ: pyrroloquinoline quinone
ALDH: aldehyde dehydrogenase
MCD: Molybdopterin cytosine dinucleotide
OCV: open-circuit voltage

References

[1] Tanino, T., Nara, Y., Tsujiguchi, T., Ohshima, T. *J. Biosci. Bioeng.* 2013,

[2] Sievers, M., Swings, J. *Bergey's Manual of Systematic Bacteriology Second Edition*; Brenner, D. J., Krieg, N. R., Staley, J. T., Garrity, G. M., *The Proteobacteria Part C The Alpha-, Beta-, Delta-, and Epsilonproteobacteria;* Springer, 2005, Vol. 2, pp. 41-95

[3] Toyama, H., Mathews, F. S., Adachi, O., Matsushita, K. *Arch. Biochem. Biophys.* 2004, 428, 10-21.

[4] Ameyama, M., Matsushita, K., Ohno, Y., Shinagawa, E., Adachi, O. *FEBS Lett.* 1981, 130, 179-183.

[5] Matsushita, K., Toyama, H., Adachi, O. *Adv. Microb. Physiol.* 1994, 36, 247-301.

[6] Chinnawirotpisan, P., Theeragool, G., Limtong, S., Toyama, H., Adachi, O., Matsushita, K., *J. Biosci. Bioeng.* 2003, 96, 564-571.

[7] Lovely, D. R. *Nat. Rev. Microbiol.* 2006, 4, 497-508.

[8] Watanabe, K. *J. Biosci. Bioeng.* 2008, 106, 528-536.

[9] Osman, M. H., Shah, A. A., Walsh, F. C. *Biosens. Bioelectron.*, 2010, 26, 953-963.

[10] Pant, D., Van Bogaert, G., Diels, L., Vanbroekhoven, K. *Bioresour. Technol.*, 2010, 101, 1533-1543.

[11] Wang, H. Y., Bernarda, A., Huang, C. Y., Lee, D. J., Chang, J. S. *Bioresour. Technol.*, 2011, 102, 235-243.

[12] Oliveira, V. B., Simões, M., Melo, L. F., Pinto, A. M. F. R. *Biochem. Eng. J.* 2013, 73, 53-64.

[13] Ikeda, T., Kano, K. *Biochimica. Biophys. Acta* 2003, 1647, 121-126.

[14] Tkac, J., Svitel, J., Vostiar, I., Navratil, M., Gemeiner, P. *Bioelectrochemistry*, 2009, 76, 53-62.

[15] Karthikeyan, R., Sathish Kumar, K., Murugesan, M., Berchmans, S., Yegnaraman, V. *Environ. Sci. Technol.* 2009, 15, 8684-8689.

[16] Rengasamy, K., Berchmans, S. *Bioresour. Technol.* 2012, 104, 388-393.

[17] Puig, S., Serra, M., Coma, M., Balaguer, M. D., Colprim, J. *Water Sci. Technol.* 2011, 64, 904-909.

[18] Yu, J., Seon, J., Park, Y., Cho, S., Lee, T. *Bioresour. Technol.* 2012, 117, 172-179.

[19] Sun, J., Hu, Y. Y., Bi, Z., Cao, Y. Q. *Bioresour. Technol.* 2009, 100, 3185-3192.

[20] Kalathil, S., Lee, J., Cho, M. H. *Nat. Biotechnol.* 2011, 29, 32-37.

[21] Feng, Y., Wang, X., Logan, B. E., Lee, H. *Appl. Microbiol. Biotechnol.* 2008, 78, 873-880.

[22] Wen, Q., Wu, Y., Zhao, L. X., Sun, Q., Kong, F. Y. *J. Zhejiang Univ. Sci. B* 2010, 11, 87-93.

[23] Lu, N., Zhou, S., Zhuang, L., Zhang, J., Ni, J. *Biochem. Eng. J.* 2009, 43, 246-251.

[24] Durruty, I., Bonanni, P. S., González, J. F., Busalmen, J. P. *Bioresour. Technol.* 2012, 105, 81-87.

Index

A

B

C

D

E

F

G

H

I

N

O

P

T

U

V

W

X

Y

Z